The Trumbly Show

A Narrative of Training Under Legendary Boatbuilder Joe Trumbly

Front cover: Boatbuilding shop at the LH Bates Vocational Technical Institute, Tacoma, 1979. Joe Trumbly is at the lower left in a white coverall, talking with another instructor; MH Dick's 14-foot sailboat under construction is at the upper left.

Joe Trumbly in his home shop, Raft Island, 1979

The Trumbly Show

A Narrative of Training Under
Legendary Boatbuilder
Joe Trumbly

Published by Matthew H. Dick
Sapporo, Japan
2022

2nd edition (first print edition), 2022
i–xiv + 272 pp.
Revised from the Kindle edition.

Printed by Lulu Publishing
Available at the Lulu Bookstore
https://lulu.com/shop

ISBN: 978-0-9910356-4-9

Published by MH Dick
dickmatthewh@gmail.com

Special discounts are available on quantity purchases by corporations, associations, educators, and others. For details, contact the publisher at the above email address.

Dedicated to

Joe Trumbly

Mark Klarich

Kathy Putney Opheim

Contents

List of illustrations

Acknowledgements

I am grateful to the following people who provided information on Joe Trumbly, the Bates Boat Building Program, or Tacoma boatbuilding through interviews (by telephone unless otherwise noted); helped track down literature or other resources; or helped in other ways, as noted. The years of interviews or assistance are in parentheses. People appearing in the text during the period 1977–79 are mentioned here only if I contacted them during or after 2002.

George Chambers, Tacoma (2002). **Patrick Cloherty**, Chandler, Arizona; high-school student in Trumbly's class 1960–61; later completed program 1969 (2004). **Chuck Graydon**, Tacoma; Bates boatbuilding instructor 1999–2012; conversations at the Bates South Campus (2003, 2004). **Kiko Johnston**, student of John Possin; conversation in Hilo, Hawaii (2005). **Barbara Keck**, Sumner, WA; Etta's daughter and Trumbly's stepdaughter (2003). **Chuck Knapp**, long interview in Boulder, Colorado (2003); Chuck also provided me with an untitled portfolio on his boatbuilding career, submitted to Naropa University to obtain one year of academic credit for life/work experience, and read an early draft of Chapters 8–12 for accuracy. **Martha M. Dick**, Taos, NM; critically read an early draft (2003). **Fred and Maggie Ogden**, Tacoma; provided logistical support (2003–04). **Katie Ogden**, Tacoma; helped with library research (2004). **Kathy Putney Opheim**, Pleasant Harbor, Alaska; discovered and mailed my photograph file, without which I would not have attempted this book (1998). **John Possin**, Tacoma; Trumbly's student 1958–59, Bates boatbuilding instructor 1979–1998 (2004). **Jeanie Robinett**, Trumbly's first wife; long interview in Freeland, WA (2003). **Sarah M. Taranto**, Grass Valley, CA; critically read and checked a late draft of the manuscript (2013). **George Thornhill**, Tacoma; Trumbly's student 1956–57 and fellow powerboat racer; he had the best long-term memory for detail I've ever encountered (2004). **Stella Trumbly,** Tacoma; Fritz Trumbly's widow (2002). **Susan Trumbly**, Gig Harbor; Trumbly's daughter (2003). **Irving M. Warner**, Kodiak; critically read parts of an early draft (2003) and stressed the necessity of getting myself into the story early on and telling it through my own eyes.

Terminology and usage

The profession I wrote about is sometimes written as "boat building" and the people who engage in it as "boat builders." I used instead the condensed forms "boatbuilding" and "boatbuilders," In fact, the usage was and still is variable; for example, Bates offered a Boat Building Program, but I worked at Knapp Boatbuilding.

The traditional convention of conferring feminine gender on boats, as in "She was a beautiful boat," is by no means universal. In fact, most professional boatbuilders I knew in Tacoma and Alaska referred to boats in neuter gender, as in "It was a beautiful boat." I tend to treat boats I like as feminine, but those I either don't know or don't like as neuter, so my own usage is variable.

I often use the pronouns "he," "him," and "his" in the strictly grammatical sense to refer to people of undesignated gender, as in "If the plug sticks to the mold, a boatbuilder has a big problem on his hands." I disclaim any sexism in this usage. Women can build boats as well as men; they proved so in the Tacoma shipyards in WWII and continue to prove so as the number of women boatbuilders increases.

Preface

This book is about the building of boats, primarily in wood but also in fiberglass. While it is intended for a general audience and requires no prior knowledge of boats, it will also be of interest to professional boatbuilders and educators.

The book is a narrative account of the 19 months I spent in the Boat Building Program at LH Bates Vocational Technical Institute in Tacoma, Washington, from 1977 to 1979, and of my work in boatyards during that period. The instructor was Joe Trumbly, who had taught the course for 22 years when I arrived and was in his last two years before retirement. At the time, he was nationally known among boatbuilders, and his training program was regarded as the best in the United States. He was an extraordinary teacher who ran his class the only way he knew—as a working boatyard, with him as foreman.

Much of Trumbly's strength in teaching derived from his unusually diverse career. He worked as a welder on steel ships during WWII and as lead man in boatyards during the last years of production wooden boatbuilding. He designed, built, and raced powerboats. He designed and single-handedly built his own 40-foot wooden sailboat (and after retirement, a 51-footer). He designed production and one-off sailboats on commission. He designed and modeled propellers, had them cast, and tested them. He invented new tools and techniques in boatbuilding and was arguably the best boat loftsman in the world.

Trumbly is not as well known today as he should be, because he devoted all his time to boatbuilding and none to writing; he published nothing himself, and relatively little has been written about him. His legacy lies almost entirely in the hundreds of students he trained in boatbuilding and the knowledge he disseminated through them.

When I started this book in 2002, I had the vague idea of writing a biography of Trumbly based on what I'd learned of him during my time in the Bates program, and on interviews of his former colleagues and students, family members, and Trumbly himself.

Two problems soon emerged. One was information. In 2003, I interviewed several members of his family living in the Puget Sound area and realized that while they knew pieces of his family life, they knew little about his professional life. Trumbly was married three times, and none of his children or stepchildren grew up continually with him. Likewise, most of the former colleagues and students I contacted had interacted with him for relatively brief periods, just as I had. His brother Fritz, a boatbuilder and at times Trumbly's business partner, had passed away in 2000.

Trumbly himself would have been the best source, but I was a couple years too late. Around 2000 he began to develop severe senile dementia and became scarcely able to recognize even his immediate family, let alone remember past events. It was a huge irony that I spent an afternoon in June 2003 looking for articles on Trumbly in back issues of *The Peninsula Gateway* in the newspaper's offices in Gig Harbor, when the man himself lived a few blocks away, at home under assisted care.

The other problem was that a straight biography would not have been very interesting, unless perhaps in the hands of a skilled fictional biographer, which I am not. Building boats was the focus of Trumbly's life. I know of no controversy or scandal, no momentous internal or external conflicts, none of the stuff that makes good grist for an exciting biography.

Nonetheless, he was an exceptional, interesting man who had a profound effect on the lives of many of his students and peers. Though I eventually ended up as a professional biologist rather than a boatbuilder, Trumbly's training program was the most intense educational experience of my life, and the things I learned from him in addition to boatbuilding have served me well—his absolute insistence on top-quality work, his approach to problem solving, his willingness to discard tradition when he found something better, his teaching style (he was by nature a showman, hence the book's title), and his passion for his field of endeavor.

In this book, I tell Trumbly's story from the point of view of my interactions with him and with the boatbuilding industry of which he was such an integral part. I have tried to include enough technical detail to give readers a sense of the complexity of boatbuilding, but hopefully not so much as to overwhelm them. A Glossary defines most of the technical terms used, and a Bibliography includes a selection of the reference texts available for people who intend to build a boat.

The book was first released in 2013 in an Amazon Kindle edition. The version here is the first print edition, somewhat revised from the electronic version. I added three paragraphs to Chapter 6; eliminated the previous Chapter 10 because it was extraneous; made corrections and added some text to newly numbered Chapter 15; updated Chapter 16; corrected some glaring errors in the Glossary; corrected misspelled, omitted, duplicated, or misused words; streamlined the text where I could; and included additional photographs. The bulk of the text remains unchanged.

Chapter 1
Beginning in Tacoma

The desire to build a house is the tired wish of a man content thenceforward with a single anchorage. The desire to build a boat is the desire of youth, unwilling yet to accept the idea of a final resting place. ... When it comes, the desire to build a boat is one of those that cannot be resisted.

—Arthur Ransome, *Racundra's First Cruise*

On a crisp day in early September 1977, I arrived at LH Bates Vocational Technical Institute in Tacoma, Washington to begin a two-year course in boatbuilding. I already had a bachelor's degree in biology from the University of Alaska and since college had earned my living as a field biologist. Friends were thus mystified by my matriculation at Bates. Several acquaintances were scornful—going to voc school, for god's sake, what a loser! In a way, I was as surprised by this turn of events as anyone, for nothing in my upbringing had predicted an interest in boatbuilding.

I grew up in the San Luis Valley of southern Colorado, a part of North America similar to the place where Odysseus settled after lugging an oar inland until he found no one who knew what it was. At 7600 feet altitude, the Valley is a high desert plain approximately 90 miles long by 50 miles wide, flatter than Kansas, hemmed in by the Sangre de Cristo Mountains to the east and the San Juan Mountains to the west. The main aquatic feature is the Rio Grande, a shallow, muddy smear snaking across the valley from northwest to southeast. You don't need a boat to cross the Rio Grande; it is perfectly navigable by wading.

When I left home for college at age 17, the entirety of my nautical experience consisted of a one-day family cruise on a chartered sport-fishing boat out of Corpus Christi, Texas; an hour paddling a canoe with my grandfather on the San Antonio River, Texas; and two frustrating half-day attempts to float down the Rio Grande in a rubber raft. I knew as much then about boats as the average cowboy knows about opera.

The decade between my leaving home and arriving at Bates Voc was a random walk leading from absolutely no interest in boats to an interest strong enough to drop everything else to learn to build them. I'd started college with an interest in birds, which led to a summer job in coastal Alaska after my freshman year. This in turn led to jobs on the Alaskan coast over the next eight years, and these jobs involved boats to a greater or lesser extent. I was at first indifferent to boats, then afraid of them, then learned to use them, and then repaired one for my employer. Finally I bought and repaired a derelict boat for myself. By this time, I'd come to feel at home on the sea, and this sparked the desire to build a sailboat and disappear into the sunset—a common fantasy that has led many dreamers, young and old, to boatbuilding.

This was how I found myself in downtown Tacoma before 8 AM on a September morning, standing in front of the main building of Bates Voc. It was a big, ugly, brick building that occupied a full city block along Yakima Avenue. Three stories tall, it was dug into the crest of a ridge so that from Yakima Avenue it looked only one story tall. The building was U-shaped, laid out around a paved courtyard that opened onto a side street off Yakima.

I entered the building and asked at Administration how to get to the boatbuilding shop. I had two choices. I could descend a stairway, cut through the electrical shop, and enter boatbuilding that way. Or I could go back outside, walk halfway around the building to the courtyard, and enter boatbuilding from the courtyard. I went through the electrical shop.

The boatbuilding classroom was a walled-off space on ground level in one corner of a large, high-ceilinged workshop. I entered the classroom and sat down at one of the long tables there. I felt rather foolish, like a first-grader on the first day of school. But my mother was not hovering outside and my schoolmates were decidedly un-firstgraderly. They were a scruffy lot wearing old clothes—canvas work pants coated with enough dried glue to stand by themselves; blue, pinstriped work shirts; scuffed boots.

At 27, I seemed to be roughly of median age for the class. Most of the students were in their twenties, but there were some downy-cheeked teenagers and two retired military men. One of the latter had a graying beard to his sternum and beefy, tattooed arms; I later learned he was an ex-Navy Chief. He looked his part, but the other—a short, balding, ex-Air Force navigator with wire-rim glasses—looked more like an accountant than a man who a few years before had flown B-52 missions over Vietnam.

The boatbuilding program had a staggered admission, so that at any given time only a few of the students were beginners. I was among four beginners that September. We got no collegial greetings or personal introductions from the other students, who'd been at Bates anywhere from a few months to nearly two years. They regarded us beginners as unformed, unknowing of anything that counted there. In a caste system founded on the ability to build a boat, they were the Brahmans and we were the Untouchables.

The old hands bantered among themselves. We beginners pretended nonchalance as we glanced around the dusty clutter of the classroom **(Illustration 1).** Hung around the walls were wooden half-models of sailboat hulls and wooden design models of propellers, in various stages of completion. Rolled, dog-eared blueprints of boat plans filled one corner. Some leaden "ducks" and a drafting spline, paraphernalia used in designing boats, lay at the end of a table. Occupying a long wall of the classroom was a blackboard. It had been cleaned over the August break, but in one corner the janitor had missed the words "Simpson's formula" and the remnants of some calculations.

An ancient, scarred wooden desk dominated the head of the room. Every square inch was covered with an assortment of ragged papers, books, open boxes of bronze and stainless-steel screws, a few large stainless-steel nuts, a length of propeller shaft, a mangled bronze propeller, bits of sailboat hardware, and an assortment of small tools, both recognizable and obscure. If a clean desk is the sign of a sick mind, then the owner of this desk was sane indeed.

The racket of a motorcycle in the courtyard outside drowned talk. Exhaust fumes drifted through open windows as the machine sputtered to a halt. A man in a heavy, brown, button-down corduroy coat propped the cycle, dismounted, and removed a motorcycle helmet. Carrying a battered briefcase, he strode purposefully through the shop and into the classroom.

This was the first time I laid eyes on Joe Trumbly, master boatbuilder and instructor. **[Frontispiece, p. ii]** He opened a tall steel cabinet behind his desk, doffed the coat, and donned a white, glue-spattered coverall over his slacks and polyester sport shirt. I sat there agape. He wasn't what I'd expected for a master boatbuilder—I'd known three boatbuilders in Alaska, and they'd all been burly, rough-looking men. In contrast, Trumbly was small and wiry, maybe five and a half feet tall, around 60 years old, with thinning white hair combed back. He had faintly walnut-colored skin, both from sun and because he was one-eighth Osage Indian. He walked and talked like a man in his full prime: no creakiness, stoop, or impaired faculties. In his white coverall he looked more like a professor emeritus of biochemistry than a working boatbuilder.

Two aspects of Trumbly's appearance stood out the first time I saw him: the eyes and the hands. The eyes were full and brown, with slight epicanthic folds. To me they looked unaccountably like the eyes of Zen masters you see in old photographs. More than intelligent, they seemed to be windows of passion, wisdom, and compassion. I later learned that the passion was for anything to do with the design and construction of boats. The wisdom went deeper than profound knowledge of a discipline. It was the wisdom of a man who had lived fully a long life and was still going strong. The compassion was that of a man who had seen his own shortcomings and those of many other people and no longer judged, but simply accepted.

Then there were the hands. They were small because the man was small but wiry and powerful from work. Remarkably, the ends of the fingers were worn down and the fingertips were splayed, without much fingernail left on some of them. These hands had handled rough materials and tools for over forty years, and they suggested the expression "fingers worn to the bone" had an origin in reality. One of the fingers was missing a joint or two.

Trumbly pulled a standard teacher's grade book from his briefcase and added the new students to his class list. He took roll and then bantered with the returning students a bit, asking how their summer break had gone, whether so-and-so had gotten a job, and so on. Finally he got down to business.

"For those of you who don't know me," he said, "my name is Joe Trumbly, and I've taught this course for 22 years. Before that, I worked in shipyards and boatyards. The minimum requirement to be hired as an instructor here is 12 years of journeyman experience. I started boatbuilding before most of you little kids were born, and while the rest of you were still in diapers." This, I came to learn, was a Trumblyism; he called all his students "little kids," teenagers and ex-Navy Chiefs alike.

"I don't have many rules here, and the rules I do have are very simple. You guys who've been here awhile know them; if you didn't, you wouldn't still be here. But it won't hurt you to hear them again.

"The first rule is attendance. You're absolutely required to be here, from 8 to 3, five days a week. You're allowed one unexcused absence per semester. Any more than that, you'll get a warning from me, and after three misses, you'll be out of here. There's a waiting list a mile long of people who want into this course. There's someone, somewhere, right now waiting to fill each of your places.

"If your mother dies or you get pneumonia, come talk to me, and we'll work it out. But I warn you, if you come to me claiming you were sick, you'd better damned well look like you've been sick. When you end up working in a boatyard, the foreman won't want to hear too much about your personal life—you're either there or you're not, and if you're not, he'll find someone else."

At this point, the door rattled open with a bump as it hit a foot. A tall, gangly young man shuffled into the room, red-faced and out of breath, but looking sleepy. With a beard and long, brown hair tied in a red bandanna, he was a flashback to the hippie era not long past.

"Is this the boatbuilding course?"

It was a superfluous question, given the boat half-models and propellers on the walls, but the guy was embarrassed and not too observant just then. The room was as quiet as the sea bottom.

"We start at 8 AM," Trumbly said, staring at him without a hint of a smile. "What's your name?"

The guy told his name, then said, "I'm late because I got in the wrong building." He started an elaboration of his meandering, but Trumbly cut him short.

"Please take a seat. I'm sorry you had trouble finding this place, but everyone else managed to. I don't like it when people show up late, so don't let it happen again."

Trumbly paused for a few beats and an impish twinkle crept into his eyes, as though to indicate his case of seriousness wasn't terminal.

"Look, I hate to get up early in the morning too, hate it as much as anyone else. I grew up on a farm in Grant's Pass, Oregon, and every day when I was young I got up at 4 AM to milk the cows. I couldn't wait to get away from there. When I finished high school, I hitchhiked to Tacoma and hired on in a shipyard. Getting up for the shipyard was a piece of cake after farming. I hated every one of those fucking cows, but if it hadn't been for them I'd probably never have gotten started in boatbuilding."

This, I later learned, was a famous Trumbly trait, throwing in bits of his life—often edited or embellished to suit the purpose at hand—with serious matters. He also peppered his speech with profanity, which I was surprised to hear coming from a teacher until I realized it was an affectation that goes with the trade. Boatbuilders are, after all, cousins to sailors. Some of the old hands grinned. They were accustomed to the Trumbly Show.

"Another rule is that no sort of roughhousing or horseplay is permitted, either in the shop or out in the yard. Just don't do it. The first time I see you at it, you'll be expelled. It's as simple as that. The shop out there is full of dangerous machinery. That big band-saw can slice through a foot of solid oak, and it wouldn't even slow down going through an arm."

Trumbly held up the hand missing part of a finger.

"I did this a few years ago, ran it into a jointer. I was pushing a board and a student was catching. Somehow I ran my finger in with the board. I didn't feel a thing, not even a tug, didn't know it had happened. I was turning back to grab another board, and the student noticed blood all over the table of the machine. He shouted at me. I saw the blood, and looked to see if the kid was hurt. He didn't seem to be. Then I happened to see my own hand, and a finger was gone. I still didn't feel it.

"So, you have to consider, if I did this with all my experience, when I wasn't tired, and when I thought I was concentrating, what can happen when a couple of guys start horsing around. They'll hurt either themselves or someone else, and I couldn't live with myself if a student got seriously injured in my shop. That's why, if I see you screwing around, I'll boot your asses out. I treat my students as adults and expect them to act accordingly."

He let this sink in for some moments, and then continued.

"There's only one other thing I want to mention. In this course, you learn the general principles of boatbuilding. I myself prefer wooden boats, but once you learn to design and build those, you can also build metal and fiberglass boats. I'm not averse to metal or fiberglass; in fact, I built steel ships during the war, and fiberglass sailboats I designed are in production right now at a yard started by some of my former students. But even if you build with these other materials, you need to work wood for patterns, molds, and finishing.

"In a boat, the main criteria are whether the vessel is seaworthy, strong, and built to last—and you want as well that the thing doesn't look like a dog. There's no single right way to build a boat. There are proven techniques, but there is also much room for innovation. If there weren't, we wouldn't have fiberglass and metal boats—hell, we'd still be using skin boats and hollow logs. I myself have invented new tools and techniques over the years, and I'll teach them to you. But that doesn't mean they're carved in stone."

At this point, Trumbly paused and asked, "Are there any questions?"

One of the new guys raised his hand.

"When I was in high-school," the guy said, "our shop teacher made us hold the hammer at the end of the handle. If he saw us holding it halfway, he'd saw the handle off and make us pay for the hammer. I'm just wondering if you do that here."

"That's the stupidest rule I've ever heard. I don't care how you hold the hammer; you can hold it with your teeth if that works for you"—Trumbly laughing here—"as long as you get the job done, and get it done right. It can't be sloppy or half-assed. If it is, you'll tear it out and re-do it, as many times as it takes to get it right."

Trumbly stopped, as though considering whether to say anything else on the matter. He decided to say more.

"Look, here's how it is. Often in your careers, you'll be tempted to do shoddy work. I know, because I've been tempted myself. You'll come in on Monday with a hangover, or you'll be thinking about your girlfriend, or the foreman will be breathing down your neck because he wants something done yesterday. But you have to consider, when a customer sees your final product, he won't know about any of these things. If it's a bad job, he'll know only that you did it, and he'll judge you by it. You'll get a bad reputation, which in boatbuilding like any other business is something you don't want. And if I recommend you for jobs in boatyards and you don't measure up, I'll get a bad reputation. So, even if you don't care about your reputations, at least care about mine!"

Trumbly decided it was time to get to work. The returning students filed out into the shop to continue whatever they'd been doing before the summer break. We beginners remained in the classroom, where Trumbly lectured until the mid-morning coffee break.

⌒

Before going upstairs to the cafeteria for coffee, I briefly explored the shop **(Illustrations 2, 10).** It was perhaps 80 feet by 40 feet in dimensions, its ceiling two stories high. In the walls facing Yakima Avenue and the courtyard were tall banks of large-pane industrial windows extending from head height to the ceiling. The shop was well illuminated by fluorescent lights and ambient light from the windows.

A second-story, L-shaped loft 15 feet wide ran along one wall and part of another. The loft had a wooden floor, and a railing that prevented people from falling off the edge. Attached to the loft, outside the railing and overhanging the shop floor, was a lumber storage rack filled with long boards. The classroom was located on the first floor under the base of the L of the loft; the only other intrusion onto the floor of the shop was a small tool-storage room under the long arm of the L.

Various pieces of equipment were scattered around the shop floor, including two 14-inch band saws, a 36-inch band saw, a table saw, a jointer, an industrial-scale planer, a wood lathe, a heavy laminating table, several wood-topped work tables, and a cube-shaped machine that looked like a giant router upside down on legs—later I learned that this was a tilting-arbor plank cutter. Particle collecting ducts ran from the machines to larger collecting ducts that threaded their way toward the ceiling like giant pythons. Scattered among the machines were strongback frames supporting small boats under construction.

In the courtyard just outside the shop was a large wooden sailboat under construction, protected overhead by a flat-roofed, lean-to shed built against the side of the building and open on the two ends. Also in the part of the courtyard claimed by boatbuilding lay a pile of scrap wood, a steam box, and some molds for fiberglass hulls. Other shops had similarly idiosyncratic materials in the courtyard—there were transformers and big electric motors outside the electrical shop next to boatbuilding, and greasy engines and some derelict cars across the courtyard in front of the automotive shop.

The shop was a semi-familiar place; I knew the names of most of the machines from my high-school woodworking class, but my understanding of them didn't go much farther. It was clearly a highly utilitarian work-space, but with cinderblock walls and a concrete floor, it struck me as ugly. My first feeling was one of disappointment. I had expected an earthier, more airy shop with a wooden floor—something less starkly industrial.

‿

When I arrived in Tacoma a week before school started, I checked into a skid-row hotel, one of many like establishments nestled near the bus station on Pacific Avenue. All I brought with me was what I could carry in my backpack. After a few days of scouring the newspaper and inspecting rentals by bus and taxi, I took an unfurnished, street-level apartment on Tacoma Avenue a mile from Bates. I furnished the place with a straight-backed chair, a table, and some kitchen utensils, all purchased at yard sales. I could now consider myself a resident of Tacoma.

The City of Tacoma occupies a peninsula shaped like a leg of mutton propped up on its base, with the foot jutting northwestward into the south end of Puget Sound. The tip of the peninsula is taken up by Point Defiance Park, roughly a square mile of woods and green. The rest of Tacoma is predominantly residential, with businesses localized in scattered neighborhood centers and along major thoroughfares. There are several university campuses and twenty or more parks of various sizes. Tacoma calls itself the City of Parks.

Tacoma is a port city's port city, bounded by the sea on two sides, Commencement Bay to the east and a passage called The Narrows to the west. A massive suspension bridge spans the Narrows. Indeed, this bridge is Tacoma's greatest claim to fame, for in 1940 its predecessor began to oscillate in a strong wind and then collapsed. Engineering texts still discuss the old Tacoma Narrows Bridge as an example of how not to build suspension bridges, and the new one as a favorable counterexample.

Downtown Tacoma lies along the lower eastern flank of the peninsula, overlooking the head of Commencement Bay. Built on the side of a ridge, the downtown is steep. It plunges like an Olympic diver for a full six blocks—from Yakima Avenue, which runs along the crest of the ridge, to Pacific Avenue a couple of blocks from the water. On the rare occasions when it snows in Tacoma, you sometimes see cars spinning like discuses out of control down steep, icy streets.

The downtown as a whole is pleasant. By the end of the 1970s, Pacific Avenue north of the 11th Street bridge was beginning an urban renewal. Some of the old buildings had been converted to chi-chi mini-malls, or at least had received facelifts. South of 11th, however, the avenue remained a scarlet woman with her skirts up, well endowed with porno shops, wino hotels, and liquor stores, and frequented by whores, bums, Gypsies, GIs, and students. Some of the Gypsies had bought empty stores in the district and moved into them. Walking down the street, you'd occasionally look through a display window expecting to see merchandise, but instead see an ornate Oriental carpet, elaborate wall hangings, and a television, with Gypsies sitting on the floor watching it. There'd be no other furniture.

The downtown waterfront had not yet seen a hint of urban renewal. A few gritty industrial buildings fronted a polluted, man-made channel called City Waterway, scattered along which were docks for boats ranging from tugboats to yachts. Separated from Pacific Avenue by a sheer drop and some railroad tracks, it was a no-nonsense, working waterfront, not the sort of place for aperitifs on cobbled terraces or midnight lovers' strolls.

Downtown Tacoma overlooks a vast industrial area called the Tideflats, or Flats. The Tideflats are so named because they were once just that, the flat tidal estuary where the Puyallup River and Hylebos Creek flowed into Commencement Bay. In the first half of the 20th century, straight, permanent outlets were dredged for river and creek, and the estuary was drained to become the greater Port of Tacoma.

Four major and several minor man-made ship channels extend like a lecher's fingers from the bay across the flats. Industries heavy and light, toxic and benign line these waterways, well supplied by seagoing shipping. There are docks, cranes, warehouses, storage tanks, marinas, boatyards, lumberyards, fiberglass fabrication shops, chemical factories, pulp mills, scrap-metal yards, and equipment manufacturers.

From downtown Tacoma, the Flats look like a war zone, shrouded in the haze of industry like smoke from a desperate battle. Incongruously, on a clear day, massive, white-sloped Mount Rainier dominates the skyline view towards the head of Commencement Bay like a giant pearl, reducing to insignificance the human bustle on the Flats.

Although it was once larger and more prosperous than Seattle, Tacoma has gained a reputation over the years as the ugly stepsister to the queen city of Puget Sound. This reputation is not entirely undeserved. When I was there, the pulp mills and chemical factories filled the air with malodorous, noxious vapors. When the wind was blowing from an easterly direction, these effluents stank up the city like Hell itself. Some progress was being made; the municipality of Ruston, nestled like a cancerous wart against Point Defiance Park, harbored a recently shut down smelter that had spewed sulfurous, arsenic-laden gases over the entire city for decades. Nonetheless, Tacoma still smelled bad enough to justify the parody sung by local wags to the tune of Goeff Stephens's Winchester Cathedral:

"Tacoma's aroma—you're gettin' me down—I gotta do sumthin'—I gotta leave town."

⟿

At first I found it difficult to adjust to city life. My apartment was a grungy dive, but I'd lived in worse. The problem was that I'd never lived in a city before and felt entirely hemmed in. For the previous two years, I'd lived in Kodiak, Alaska, where you could reach the edge of wilderness within a 15-minute walk from any part of town.

What preserved my sanity during my first few months in Tacoma was Wright Park. Occupying an area of two by five city blocks located directly between my apartment and Bates, Wright Park was a botanical garden. It had an antique greenhouse full of exotics, a pond with ducks, and trees from all over the world labeled to genus and species. For the first few months, I left early for school so that I could dawdle my way through the park, feeding bread to the ducks and imagining I lived not far from wilderness. Later, when I got acclimated, I dashed through the park in the morning so I wouldn't be late for Trumbly's roll call.

Aside from another student from Kodiak, I initially knew no one in the city. Two weeks into the fall, one of my classmates, Mike, held a keel-pouring party. He'd completed the Bates program the previous July, but Trumbly was letting him continue a few weeks into the fall to wrap up a project.

Mike was ready to start building a cruising sailboat in his backyard in South Tacoma, and he'd constructed a mold for the vessel's lead keel. The mold was a tapering box made of oak strips, with part of the top open. An initial step in the construction of a sailboat, a keel pouring is not only practical but also ceremonial, like laying the cornerstone of a building— hence the party.

One Saturday morning we gathered at Mike's place in our work clothes. Maybe half the boatbuilding class was there, along with some girlfriends and a few alumni, former classmates of Mike. We stood around a makeshift crucible fashioned from a length of two-foot-diameter steel pipe. We sipped beer as we watched a big propane torch melt half a ton of scavenged lead in the crucible. When the lead had melted, Mike skimmed the slag off the top and then pulled the stopper from a length of one-inch pipe welded into the base of the crucible. The molten lead gleamed like quicksilver as it flowed into the mold, from which acrid smoke rose as the inner surface charred (**Illustration 3**).

The keel pouring wasn't very exciting as a party, but it was the strangest reason I'd ever encountered for one. It was also a turning point for me. I didn't feel alone in the city anymore.

Chapter 2
Origins of an interest

When I had [the ship model] *in my hands I felt as though the breath had been knocked out of me. ... I had never really seen a ship before. The large floating shapes were not part of my life, did not move me. But the model, held in my hand, made it somehow possible for me to comprehend.*

—Don Berry, *To Build a Ship*

LH Bates Vocational Technical Institute began in 1941 as the Tacoma Vocational School to fill a demand for trained workers created by the burgeoning war effort. By 1977 it was a big enterprise, one of the largest vocational schools in the country. The campus occupied two full city blocks in downtown Tacoma, only a block above the public library. Boat Building was just one among 50 training programs the school offered, spanning the alphabet from Accounting to Welding. I don't know if they had Zookeeping, but they might have.

Though it was thus only a small part of Bates, Boat Building nonetheless had special status. Over nearly a quarter century, Trumbly had steadily built a national reputation for the program. Admittedly there wasn't a lot of competition then, for only a few other institutions in the US offered intensive training in boatbuilding. Edison Technical College in Seattle had begun a boatbuilding program in 1936. Later Edison became amalgamated with Seattle Community College, but around Puget Sound the boatbuilding program was still widely referred to as "Edison." On the East Coast, serious training in boatbuilding was available at a place called The Apprenticeshop in Bath, Maine—a mere upstart founded in 1972 by a guy named Lance Lee. I learned about the Apprenticeshop the first fall I was at Bates, when Lance Lee spent a few days in our shop observing how

Trumbly conducted training.

A salient feature of Bates was that it was cheap. Bates was part of the Tacoma School District, and as such was publicly funded. In 1977, the tuition was around $750 for an 11-month school year, whether or not you were a resident of Washington State. I felt almost guilty, getting the best training money could buy at bargain-basement cost. Tacoma must have felt it was making out, though, because many Bates graduates remained in the city and fed their paychecks into the local economy for the rest of their lives.

We boatbuilders believed ourselves to be the elite of the institution and viewed the other programs with disdain. The automotive geeks working directly across the courtyard seemed to us a primitive lot, closely allied with the australopithecines. They wore greasy T-shirts with cut-off sleeves in summer, leather jackets in winter. Their shop stank of oil, solvents, and exhaust, whereas ours was redolent of the subtle odors of planed cedars. Like so many trained chimps, they merely disassembled and reassembled metal objects, clearly inferior to artists in wood like us. Nonetheless, we treated them with the condescending kindness one displays toward disadvantaged children, because we occasionally needed them to weld things for us.

Then there were the poor sods in the culinary program upstairs. Bates had an active cafeteria where students did all the cooking and where you could buy doughnuts and coffee, or a palatable breakfast or lunch. We boatbuilders viewed the culinary students as unfortunates who had reached the natural limit of their depauperate abilities. Every once in a while, as though to support our view, some wretch would mistakenly put salt instead of sugar into the doughnuts. At least, we liked to believe it was by mistake.

We considered the machinists to be approximately our peers. They fashioned complex objects in metal, just as we did in wood. They cheated, however. They made settings on big, complex machines and let the machines do their carving for them.

⌐

I often wondered why a person chose one course of training over another. I could imagine various routes to a particular choice. A high-school counselor might have noted that a student had little interest in academics and on the basis of an aptitude test recommended he become a mechanic, boatbuilder, cook, or machinist. A person's parents might have told him he was too stupid for college and had better enter a trade, any trade, by whatever program was available at the moment. A person might through some random exposure to an occupation—it could have been by watching an uncle at work or bumbling into a construction site—have thought, "Gee, that's interesting. I think I'd like to do that." It was by this last route that I came to boatbuilding.

The first boat I ever really looked at was a loathsome little rat canoe. I encountered it the summer of 1968, after my freshman year in college, when I worked as a field assistant for a Dartmouth College professor named Dick Holmes. Dick was studying the ecology of sandpipers along the Kolomak River, at the edge of the Askinuk Mountains on Alaska's Bering Sea coast. Three of us—Dick, his graduate student Craig Black, and I—arrived by ski plane in mid-May before the snow melted and camped for the next 10 weeks on a rise not far from the river. The Kolomak, a tidal river 75 yards across running through low coastal tundra, could become quite choppy under the influence of tide and wind. It wasn't something we wanted to swim across, especially with chunks of ice floating in it from spring breakup.

One of Dick's study plots was on the opposite side of the river from our camp, and our only means of crossing the river was a rat canoe. The canoe was 10 feet long and about 2 feet wide. It had a flat bottom, a straight sheer, and nearly vertical sides maybe a foot high. Constructed of plywood covered with rough fiberglass and painted green, it looked like a floating coffin with pointed ends.

Dick was a big man, 6 feet tall and over 200 pounds. When he kneeled in the rat canoe—which he assured us was the correct rowing position—his high center of gravity increased the instability inherent in the boat itself to suicidal levels. Whenever Dick paddled across the Kolomak, Craig and I would stand on the bank and wager as to whether or not he'd make it. The odds varied with the velocity of the wind and the amount of chop on the river.

Against all logic, I never saw Dick capsize. Though he looked as unstable as a drunken tightrope walker, using the paddle as much for a balancing rod as for its intended purpose, he invariably crossed the river. He grudgingly admitted, however, that his proficiency had come at the expense of several plunges in the icy water.

I eventually met the builder of the rat canoe. He was Cal Lensink, manager of the Clarence Rhode National Wildlife Range, within which lay our study area. Cal was a tall, rail-thin, pipe-smoking, Dutch-American game biologist who was completely at home in the Alaskan wilderness.

He'd been an Army MP in WWII, so he was a lot tougher than he looked. He had a PhD in wildlife biology from Purdue University, read the journal Science for entertainment, and loved classical music. Cal was also a pilot; after the ice left the Kolomak, he delivered supplies and mail to us every few weeks in a Cessna 180 floatplane. On one of his visits, not knowing he'd built the rat canoe, I commented about the boat being a death trap.

Cal looked at me as though I were an imbecile. "It's only a death trap," he said dryly, "if you don't learn how to use it."

Then perhaps realizing I was young and therefore probably ignorant, Cal explained.

"Rat canoes have a specific purpose. I learned to build them in Fort Yukon. The Indians there use them for trapping, mostly muskrats, which is why they call them rat canoes. On the Yukon Flats there are thousands of small, shallow lakes, and you have to portage often. Rat canoes are light enough that you can lift one easily onto your shoulders over your backpack and portage it for miles, if you have to. They're also small and light enough to be strapped onto the float of an airplane and moved around—which is why I use them here."

This was the first time I realized there was a connection between the form of a boat and its specific function. This connection is obvious, but I simply hadn't thought about it before. I hadn't thought anything at all about boats before.

After Cal's explanation, I tried out the rat canoe on one of the few calm, sunny days we had that summer. I paddled successfully halfway across the river, then back to shore. Eventually I came to use the canoe on a fairly regular basis, but I always had a feeling of imminent disaster. Despite their particular advantages, I still think of rat canoes as plug-ugly, vicious little things.

Cal overlooked my distaste for rat canoes and hired me the next summer as a field assistant at the Old Chevak research station on the Yukon-Kuskokwim Delta. Laid down by the Yukon River to the north and the Kuskokwim River to the south, the Delta is a mostly lowland area of treeless coastal tundra and wetland, about the size of Vermont, New Hampshire, and Massachusetts combined. It is traversed by numerous rivers, between which lie countless small lakes, sloughs, and marshes. Cal told me you could travel by small boat to within five miles of anywhere on the Delta.

My field partner was Eddie Kootuk, an Eskimo about my age and a classmate at the university. Eddie grew up in St. Michael on the Yukon River but had relatives in Chevak, the nearest village to the field station. Eddie hadn't learned to drive a car, but he'd run boats since he was a boy. Our main job was to band ducks and geese. All day long, for weeks on end, we cruised along rivers and sloughs looking for flocks of flightless, molting waterfowl and their nearly fledged young. When we spotted a flock, Eddie would ground the bow of the skiff in the mud of the shore. Weighted down by hip boots and life jackets, we'd sprint off over the soggy tundra after the birds.

This was the first time I spent an appreciable amount of time in small boats. Although I itched to learn to drive one, Eddie wouldn't let me anywhere near the helm. In fact, his major source of entertainment that summer was to try to topple me head over heels whenever he could, which was often until I got my boat legs.

I had my revenge, though. One day we were chugging down the Kashunuk River in the refuge's freight boat, a heavy, flat-bottomed, 20-foot skiff. Eddie went to sleep at the throttle, and rather than wake him up, I smugly let him veer off into the riverbank at full speed. The tide was out, exposing high, sloping banks of soft mud. The freight boat plunged into one bank like a fly ball hitting a tub of chocolate mousse, then slid up it like a surfboard in reverse. The shaft of the outboard kicked up, and the spinning propeller coated Eddie with mud. Seconds later, the boat's deceleration tossed him headfirst into the bilge. Kids having fun in boats.

"You sonofabitch," Eddie shouted, "why didn't you wake me up?"

"Hey, you're supposed to be the hotshot boatman," I said. "Maybe you should let me drive from now on."

Eddie's proficiency as a boatman shone brightest on our seal hunts. As an Alaskan Native, Eddie could legally hunt seals; as a non-Native, I was technically just along for the ride. Although we shouldn't have done it on government time in a government boat, we nonetheless went after every seal we saw. To ignore a seal when he had the means to pursue it was as alien a concept to Eddie as bypassing a Macy's sale would be to a Manhattan matron.

The seals on the Kashunuk were wary. The unwary ones didn't last long, because many Eskimos traveled the river, and they all had the same attitude toward seals as Eddie. Shooting a seal was the Delta equivalent of going to the supermarket and stocking up on a week's supply of meat and cooking oil, with a pair of boots thrown in as a bonus.

When Eddie spotted a seal poking its head up for air, he'd get as close as possible to it, trying to predict where it would emerge next. He'd stand with his .222 rifle at ready, steering the boat in the wind- and tide-generated chop of the river with the outboard handle between his legs. When the seal reappeared, bobbing in and out of sight up to 50 yards away, Eddie would somehow manage to squeeze off a shot. He not infrequently hit his target in its cantaloupe-sized head. Then there was a full-speed run to try to get a harpoon in the beast before it sank. Curiously, whereas Eddie was a crack shot against a bobbing target from a tossing platform, he was a piss-poor shot on land. His well-honed compensation mechanisms simply couldn't adjust to the lack of motion.

There were no roads where we worked, and it was impossible walk very far in any direction without encountering a body of water. Therefore boats were our sole mode of transportation. The refuge had three boats, all powered by outboards. One was the freight skiff. It was relatively slow, and we used it for hauling heavy loads such as lumber and drums of gasoline. Built locally, the big skiff was similar to hundreds of others used by Eskimo families that dispersed in summer from their villages to campsites all over the Delta to drink tea, pick berries, and catch fish for drying.

Another of our boats was a Grumman aluminum canoe. With a 25 HP motor mounted on the vertical transom, the Grumman screamed over the water like a low-flying airplane. We used the Grumman to travel fast and light during banding forays. Both the freight skiff and the canoe were completely open.

Our third boat was a 16-foot, carvel-planked runabout. With a steep V forward and a round bottom, the runabout was the easiest riding of our boats. It had a covered bow for dry stowage and a windshield that deflected spray. We used the runabout for long trips, especially in bad weather. To the deck beam at the front of the cockpit was screwed a brass plaque that read:

B & B Boat Works
Marysville, Washington

We thus referred to the runabout as the "B and B." I spent a lot of time in the B and B that summer. I came to marvel at the craftsmanship required to meld hundreds of pieces of accurately cut wood, hardly any two alike, into a sturdy, watertight, functional shape of compound curves. At that time, it was the farthest thing from my mind that I might ever learn to build such a thing, but the seed of interest that would land me at Bates became firmly planted then.

⌇

Six years later, in 1975, I took a job with the US Fish and Wildlife Service to look for seabird colonies around Kodiak Island—the wild, mountainous home of the famous Kodiak brown bear—in the western Gulf of Alaska. This was during the period of federally funded Outer Continental Shelf, or OCS, research, the purpose of which was to assess Alaska's coastal resources before they got screwed up by overfishing and oil spills. The funding for this work seemed practically unlimited. The head office was in Anchorage and my boss was once again Cal Lensink.

When I arrived in Kodiak in May, my first task was to purchase a boat and outboard motor. I was still encumbered by the misapprehension that government employees should endeavor, wherever possible, to conserve taxpayer dollars. Rather than spend profligately, I purchased a damaged 19-foot Bartender skiff for $600 at a state auction. The Bartender was a double-ended, V-bottomed plywood boat with an inboard well at the stern for an outboard motor. The boat had a cabin forward, with a windshield on top to deflect spray.

A marine architect named George Calkins had designed several lengths of Bartenders. The name Bartender has nothing to do with serving drinks, but instead refers to the design's seaworthiness in exceptionally rough waters, such as those encountered over shallow bars at the mouths of large rivers.

Unfortunately, this particular Bartender wasn't too seaworthy when I bought it. There were several saucer-sized holes in the plywood of the hull, and some of the frames were cracked. No matter, I thought. I would patch it up as good as new. This was my first foray into anything to do with boatbuilding, of which boat repair can be considered a sub-discipline. I consulted local experts on how best to patch the holes and eventually arrived at the solution of covering them with plywood patches screwed to the inside of the hull. I then filled the holes on the outside with copious amounts of Marine-Tex, a black putty that hardens like rock. Marine-Tex is the nautical equivalent of duct tape.

My efforts were only partly successful. The holes I patched remained tightly sealed, but I had neglected to reinforce the cracked frames. As a consequence, working of the frames opened hidden leaks along the sides and keel that necessitated frequent bailing. Late in the season, when my field partner and I were motoring blissfully down Narrow Strait, a long section of seam opened along the keel, and we suddenly found ourselves shin deep in water. Only through Herculean efforts at bailing did we make it back to Kodiak Harbor, just barely. That ended the season.

∽

The following season, I returned to Kodiak with carte blanche to buy whatever boat I needed to do the job without exposing the federal government to a wrongful death lawsuit. The Feds would undoubtedly have paid for a new Boston Whaler, the nautical equivalent of Tupperware, one of those unsinkable modern marvels made of foam sandwiched between fiberglass. The only way you can sink a Boston Whaler is to burn it. A Whaler, however, didn't fit my increasingly romantic notion of what a boat should be.

I soon learned that in Kodiak, if you wanted a safe, sturdy wooden workboat, you bought an "Opheim," the general name referring to any boat built by Ed Opheim **(Illustration 4)**. Opheim boats had so much wood in them that if you holed one, it might fill with water but would stay afloat, outboard and all. You'd just be standing on a floating platform with water up to your knees.

Ed lived at Pleasant Harbor on Spruce Island, located across Narrow Strait from the town of Kodiak. Telephone service didn't extend to Spruce Island, and the only way to do business with Ed was by mail or in person. I was anxious to get a boat, so I decided to see him in person. I rode the mail plane from Kodiak to the village of Ouzinkie at the west end of the island and from there walked a trail three miles through tall Sitka spruce woods to get to Ed's place.

A quarter mile from Pleasant Harbor, I encountered by chance Ed's grandsons, Craig and Chris—four and six years old, respectively—who were out gallivanting in their back yard. Their back yard was essentially the entire 12 square miles of Spruce Island. The boys offered to guide me to their grandfather's shop, and as we walked along, they chattered in the most remarkably precocious display of coarse language I've ever heard. When I remarked, for example, that I wanted to purchase an Opheim dory, the four-year-old said,

"Yep, them bitch-humping mother bastards are the best son-of-a-whoring boats any fucker can buy."

He was trying to shock me—clearly, I was a source of entertainment—but I wasn't so much shocked as impressed, for Craig was a young Mozart of profanity. "Where'd you learn that kind of language? At school?" I asked.

"No, we don't go to a school. Our Grandma Anna teaches us at her house," Craig said.

"Does your Grandma let you talk like that?"

"No," Chris said. "She puts soap in our mouths when she hears us."

"So, where'd you learn to swear?"

"From our dad and our uncles," Chris said.

"But not our grandpa," Craig said. "He don't swear much, and he don't like it when we swear either. But at least he don't put soap in our mouths."

So I asked them about their dad and their uncles. Their dad Chris was a fisherman and cannery operator. Uncle David was a fisherman and outboard mechanic. Uncles Norman and Edwin were fishermen and boatbuilders. It just went to show the age at which maritime training started around Kodiak. The boys had obviously spent their formative years in the company of men who worked on the sea, knew boats, and swore like sailors, which they were.

We finally emerged from the dense woods onto the beach at Pleasant Harbor, which was a bustling little place. In the vicinity of the beach were a small-scale fish-processing plant on pilings, sheds of various sizes, and a small sawmill. A dirt road led up the hillside overlooking the little bay, and scattered along this road were a handful of tarpaper-covered shacks used to house workers for the processing plant. Ed and his sons had built all of this.

As Chris and Craig led me across the beach to the road, I noticed a miniature boat graveyard. Tossed helter-skelter alongside a building were fifty or more painted, wooden toy boats of various sizes and designs, the oldest among them beginning to rot. Ed had made them for his grandchildren over the span of a decade and a half. I wished I could take them with me, for I thought they belonged in a museum.

Ed had built his house and boat shop atop the hill overlooking Pleasant Harbor. It was a steep climb, but in 1962 a tidal wave had destroyed everything he owned down on the beach, and he'd rebuilt on higher ground. The hilltop wasn't as convenient as the beach, but it provided a superior view. To the south and southwest were the steep emerald slopes of Kodiak Island across Narrow Strait; to the southeast, a view of the Gulf of Alaska as far as the eye could see; on the nearby steep flanks of Spruce Island itself, dense, black-green stands of Sitka Spruce interspersed with light-green alder thickets and still-brown meadows.

The boys led me to Ed's shop. Built on low pilings, it was constructed entirely from rough spruce lumber sawn in the mill down by the beach. It had a peaked roof to deflect heavy snow.

Around 6 feet tall, Ed was a big man of Irish-Norwegian-Aleut descent. In his sixties, he had a full head of thick, graying black hair. Like many of the locals, he didn't volunteer information to strangers. He gave short and concise answers to my questions, and in turn asked a few penetrating questions to find out how serious I was about a boat.

As we talked I couldn't keep my eyes from straying around the shop. It was the first boat shop I'd ever seen. Light shone in through a bank of paned windows in the south-facing wall. Workbenches ran along a couple of the walls, and power machines were scattered around the floor—band saw, table saw, jointer, planer, drill press, all run by a diesel generator in a nearby shed. Small tools lay everywhere—Jorgensen clamps, pipe clamps, C-clamps, hand planes, electric drill, skill saw, brace and bit, hand power-planer, adzes, chisels, a slick, mallets, hammers, squares, tape measures, bevel gauges. Everything was well worn and marked with splashes of glue and paint. On the walls hung numerous plywood patterns for the bottoms, stems, and transoms of various sizes and styles of boats

Filling most of the shop was a massive 24-foot Opheim dory under construction for someone on the Alaska Peninsula far to the southwest. It sat upright on sturdy sawhorses, the frames braced to the ceiling joists with two-by-fours to keep them from shifting during planking. The half-completed boat looked like a huge wooden insect, crawling across the floor or hanging from the ceiling, depending on how you looked at it, leaving droppings of sawdust and shavings.

Ed built boats to order, finishing approximately one a month. He worked on a first-come, first-served basis. You paid a down payment, which would cover materials, and then waited for your boat. Ed used whatever materials you were willing to pay for. For frames he preferred the local Sitka spruce, which is light and tough. He used spruce for planking as well on the boats he built for his own use, but most customers preferred mahogany or fir for planking.

If you put money down for a boat and later decided you didn't want it after all, Ed would refund your money as soon as he sold the boat to someone else, but he'd lose some respect for you. Sometimes, when he had a little money ahead, he'd build a boat on speculation. His boats always sold, one way or another.

It turned out I might not have to wait for a boat, because Ed had a dory **(Illustration 5)** left from a cancelled order. We walked a hundred yards from his shop to a meadow where four or five boats were propped up on blocks. Products of a winter's work, the boats awaited their owners. Most of the hulls were painted white on the outside and blue on the inside. The dory was the exception, coated inside and out with a mixture of pine tar and linseed oil to give it a natural finish. It was strange seeing those boats in the meadow far above the water. From a distance, they looked like a fairy ring of giant mushrooms emerging from the dried grass.

We made the deal then and there, though the negotiations were strained. Ed didn't like doing business with the government.

"For some reason, government agencies usually want a boat in a hurry," he said, "but when it comes to paying, they take their good sweet time about it. I've had to wait as much as six months for a check, and I just can't afford that." We finally agreed I'd return with a cashier's check as soon as I could and pick up the boat then.

It took two weeks to get a check from the Anchorage office. When I returned to the hill above Pleasant Harbor, I was struck by how peaceful the place was. In the light of a sunny May day, the woods, meadows, strait, and sky stood out in hard colors, like a dream. The stillness of the cool air was broken only by the soothing dubba-dubba-dubba of Ed's generator. Ed was fortunate indeed. He had a useful occupation and a place to practice it according to his own schedule, with his family settled around him in a land tourists paid thousands of dollars to see from cruise ships.

For the next two months, my field partner Jay Nelson and I steered the dory along the northern and western shores of Kodiak Island and circumnavigated Afognak and Shuyak Islands. To lend an air of credence to our nautical abilities, we donned the white caps traditionally worn by old halibut fishermen, though we fooled none of the locals with this affectation. Indeed, we made complete asses of ourselves.

Our survival had little to do with our nautical skills. It was the dory that carried us safely past the horrific rip tides of Whale Passage and Shuyak Strait, the dory that compensated for our inability to predict the williwaws that struck like cobras. There were few mistakes we could have made so foolish that the boat wouldn't have kept us afloat and alive. With its high stem, high narrow transom, and flared sides, Ed Opheim's dory was virtually idiot proof.

⁓

Travelling for the government, I saw a lot of interesting places that I wanted to return to. Soon after the field season ended in August 1976, I decided I'd spend the following summer beachcombing and sightseeing around the Kodiak archipelago in my own boat. Basically, I wanted a Jack Kerouac On the Road experience, but Alaskan style in a boat. Since I didn't have a boat, I began looking for one I could afford. During this search, I met Ray Tufts, who above all else was responsible for my arriving at Bates the following year.

Ray was a tall, solidly built, mild-mannered, ex-college basketball player who for a decade had taught industrial arts at Kodiak High School. Finally growing weary of school bureaucracy, he'd set out on his own as a boatbuilder. He had a shop in the thick woods near Mill Bay, a few miles from downtown Kodiak. Ray filled a local market for durable plywood work skiffs he sold as the "Tuffy" line of boats.

None of the Tuffies was what I wanted, but Ray gave me an idea. He told me he'd served a brief apprenticeship under Ed Opheim several years before. He'd arranged this by ordering a skiff from Ed and then asking if he could help build it. Ed had agreed.

So I went to Ed with the same proposition, telling him I wanted to become a boatbuilder, just as Ray had. This wasn't entirely a lie; I was by then intrigued by boatbuilding. To my great disappointment, however, Ed wanted no part of it. At first I was angry, thinking he was selfishly guarding his secrets from possible future competition. That didn't make any sense, though, because he'd taught Ray, and Ray was in direct competition with him. Later I realized it was because Ed viewed me as a starry-eyed dilettante, an assessment that wasn't far off the mark. He simply didn't want to waste his time on me.

Ray and I became casual friends, and I stopped by his shop every so often to see what he was building. He had an apprentice working with him, a teenager of Hawaiian descent named Patrick Chapman. One day when I visited Ray, I caught him in the midst of trying to convince Patrick to make a career of boatbuilding. Ray was describing the Bates Boat Building program, telling Patrick how he'd spent a summer there as a special student learning design and lofting from a phenomenal instructor named Joe Trumbly. Ray had some Bates application forms, and he handed one to Patrick.

"Why don't you just fill out an application and send it in? There's a waiting list, so you don't have to decide right now whether you want to commit to two years. When you get accepted, you can decide then."

Ray then turned to me. "I have an extra form. Why don't you apply too? You're always looking at boats; you must have some interest in boatbuilding."

And so I did. I filled out the application and mailed it. I don't remember why; I certainly had no intention of spending two years at a vocational school. Maybe it was to support Patrick, who was wavering.

‿

One October day, my friend Fred Ogden told me he'd found the perfect boat for me. We went down to the harbor, where the Sargent brothers were selling a gitney for $1500. They'd had a smashingly good fishing season and wanted to upgrade their vessel.

A gitney was a classic style of salmon-seining boat used since the 1920s around central and western Alaska. It was flat bottomed, with a shallow draft for near-shore fishing. Though there was a small cabin forward, gitneys were built for work rather than comfort. The engine, for example, took up so much space in the cabin that there was barely room for the swing-down bunks. Cooking was done on a Coleman stove atop the engine cover. The open working deck comprised fully two-thirds the length of the vessel, with a seine platform aft and a fish hold in the center, accessible by the removal of deck boards. I describe gitneys in past tense, but even today there are probably still some masochists using them somewhere in Alaska.

The gitney for sale was named *Devil's Paw* **(Illustration 6)**. It was 19 feet long and nearly half that in beam, powered by an inboard Gray Marine gasoline engine as old as the boat itself. Nowhere as sleek as an Opheim dory, *Devil's Paw* wasn't at all what I'd had in mind for summer cruising. It wasn't nearly as safe, either, for if you holed the hull, the heavy engine would drag the boat to the bottom like a sinking anvil.

Because they intended to sell the boat, the Sargent brothers hadn't bothered much with cleaning or maintenance. The hold was three inches deep in fish gurry, a suspension of slime, guts, and scales with a vulture-choking stench. Although the fisherman around Kodiak joke that this is the smell of money, unlike money you can't get rid of it. Looking around the boat while trying not to breathe too deeply, I found pockets of obvious rot. The transom was about to detach from the rest of the vessel, and it was a small miracle the Sargent brothers had survived the trip back to Kodiak.

Fred was adamant. "Look, you don't want to spend the whole summer camping on the beach. You need a boat you can live on. This is a great platform; it just needs a little cleaning and some minor repair. I'll help you with the engine; we'll get it purring like a kitten, you'll see." Fred was a well driller, and he knew about engines. So I bought the gitney.

I had the boat hauled to Fred's place on Spruce Cape and blocked it up alongside his shop. I built a makeshift frame over it and covered that with visqueen plastic. Fred's sales pitch about cleaning and minor repairs proved to be one of the greatest understatements of the 20th century. That winter, I devoted many evenings and virtually every weekend to *Devil's Paw*. I spent hundreds of hours scraping, painting, and caulking in the raw Kodiak weather. I did some rebuilding, too, and in the process first encountered one of the classic problems faced by boatbuilders, that of the bevel.

What I wanted to do was add some frame members to the inside of the transom to reinforce it. This would have been simple had the transom been rectangular, vertical, and square with the bottom and sides of the boat, like the wall of a house. I could have bolted some two-by-fours inboard around the periphery of the transom, nailed the hull planking to the two-by-fours, and the job would have been done. This, in fact, was exactly what I tried to do at first, but it didn't work.

The problem was that the transom itself was tilted aft so that the bottom planking met it at an angle. Furthermore, the sides of the boat curved inward slightly toward the transom and thus also met it at an angle. Therefore, any reinforcing members inside the transom needed to be cut at the appropriate angle—or bevel, as it is called in boatbuilding—to allow the planking to lie against them with no gaps.

I am embarrassed to say that the transom stumped me for a week. I finally realized the reinforcements I was putting in had to be beveled, but when I cut them out and beveled them with a hand plane, they'd be too long or too short, or the bevel would be wrong. There are simple techniques for doing this sort of work; had I known them, I could have dispatched the transom in a few hours. But as I didn't know them, my sole strategy was trial and error.

The following spring, the time came to transport *Devil's Paw* from Fred's place back to the harbor. For this, I purchased a decrepit flatbed truck from Smoky Stover's junkyard. Fred and I worked weekends for a month, cutting and welding the frame and suspension of the truck to convert it into a massive boat trailer. Then we settled *Devil's Paw* onto a cradle of six-by-eight timbers bolted to the trailer.

For the actual launching we needed a mobile cradle, or travel lift, because the boat was too high on the trailer to use a launching ramp. A curmudgeon named T.T. Fuller owned the only mobile cradle in Kodiak. Since T.T. had a monopoly, he felt it his privilege to order and humiliate his customers, and even the most hardened fishing captains had to take this abuse. If a captain talked back, T.T. would drop his boat where it happened to be and then start charging him storage. Fred and I were not exempt; T.T. had us in his clutches for half a day.

"Move those chocks," he'd say, "—no, not there, you dumb bastards! How old are you? You're the two stupidest people your age I've ever encountered, not worth a rat's ass." And so on.

We got into a kind of Stockholm syndrome with T.T. Whenever he slipped and uttered a word of encouragement by mistake, or even a phrase without abuse, we were so grateful we wanted to kiss his boots.

I never did take a trip in *Devil's Paw*. Freshly painted gray and black, with red trim, the boat looked splendid, though it still reeked of fish gurry. The hull was watertight, but the Gray engine had various mechanical problems, and replacement parts were scarce. When Fred finally got the engine purring like a kitten, there was a blockage in the keel cooler, so the engine overheated rapidly. We managed to clear the blockage, but the engine overheated nonetheless.

Devil's Paw was still languishing in Kodiak Harbor when, around mid-July, a letter arrived from Bates informing me I'd been admitted to the boatbuilding program and was scheduled to start in September. The letter was a shock, for I'd forgotten about my application. I rushed to Ray Tufts to tell him the news and learned Patrick Chapman had also been accepted. Patrick was elated. By this time, Ray'd been telling him stories about Bates and the boatbuilding instructor, Trumbly, for nearly a year, and Patrick was packing his bags for Tacoma.

Though I went through the motions of thinking about it for a few days, I'd decided I would attend Bates immediately after I read the letter. What I still don't completely understand is why I made this decision. It wasn't that I needed a profession; I already had a profession. Nor can I claim I felt a warm glow whenever I thought about boats. I liked boats, but I appreciated them as tools rather than as objects of adoration.

As incredible as it seems to me now, I was willing to uproot myself to a strange city for two years of a student's poverty, all on the basis of a road-less-traveled intuition. I already knew the approximate subset of destinations along the path I was on. Suddenly a new path opened up, and I wanted to see where that one led. In other words, I went to Bates on a whim. I rationalized this by telling myself I would learn enough to build myself a cruising sailboat, thereby saving a great deal of money over buying one. This would justify, in practical, dollar-and-cent terms, spending two years in boatbuilding school.

I departed for Tacoma in late August, leaving *Devil's Paw* for Fred to sell as a penance for his having convinced me to buy it in the first place. He finally sold it, but overall the business transaction was not the type recommended by the *Wall Street Journal*. I'd bought the boat for $1500, worked hundreds of cold hours on it, spent hundreds of dollars in marine supplies, and made Fred miserable in the process. In the end, all I got for *Devil's Paw* was $1500. I felt lucky to get that much and sorry for the new owner.

Chapter 3
Initiation by lofting

The day was to come in the 19th century when there appeared a strong body of opinion against those smaller builders who designed by carving a model.

—Douglas Phillips-Birt, *The Building of Boats*

My first day at Bates, I had expected to start immediately working in the shop, learning exciting and manly things like planking and caulking, but this was not Trumbly's way. He required all beginners first to learn lofting, which is the initial step in the process by which the plan of a boat hull—the design, or lines—is translated into an actual hull.

Lofting performed the same function in the Bates boatbuilding program as dissecting cadavers does in medical training. Conducted in the sweltering, low-ceilinged loft (hence the word lofting) above the shop, it was hot, tedious work that had weeded out from the program more than one under-motivated student. Yet, just as anatomy is fundamental to training in medicine, so is lofting fundamental to training in boatbuilding. It was through lofting that Trumbly taught the gross anatomy of a boat. Until we finished the lofting assignment, which turned out to involve more than just lofting, Trumbly allowed us to do nothing else. There was no touchy-feely working in pairs. Each of us succeeded or failed alone.

In lofting, a boat's lines are drawn full size on a whitewashed wooden floor **(Illustration 7)**. The floor is wooden because the loftsman needs to tack down battens to draw long, curved lines with a pencil. It is whitewashed so that the lines will be visible. And full size means just that. If, for instance, a 50-foot sailboat is to be lofted, the lofting floor must be somewhat longer than 50 feet, and somewhat wider than the boat is high from keel to deck. Obviously, a shop building large enough to accommodate the construction of a 50-foot boat will have a floor large enough to loft it.

Trumbly gave each of us beginners a photocopy of the lines **(Illustration 8)** he'd drawn for a seven-and-a-half-foot, round-bottomed dinghy **(Illustration 46)**, intentionally small enough to be lofted on a 4-foot by 8-foot sheet of plywood. The lines had been copied onto a sheet of legal-sized paper, with the table of offsets—key measurements derived from the lines—left blank. The first step in the lofting exercise was to fill in the table by measuring the offsets from the lines, using an architect's rule. Each of us then whitewashed a sheet of plywood to cover a previous student's loft on the same sheet, laid it across a pair of sawhorses, and went to work **(Illustration 10)**.

⌐⌐

To fully understand the purpose of lofting, it is necessary to understand the modern method by which wooden boats and ships are constructed, which is called skeleton construction. Skeleton construction involves first erecting the rigid skeleton **(Illustration 9a)** of a vessel and then fastening the planking to this skeleton. The main elements of the skeleton are the backbone (stem, keel, stern timbers, and transom) and the frames, which are called "ribs" by nearly everyone except boatbuilders. The frames are attached to the backbone in the same orientation that ribs are attached to the vertebral column of a human skeleton, hence the metaphorical terminology. Other skeletal elements such as floor timbers, deck beams, and long stringers running fore-and-aft inside the frames help strengthen the skeleton.

The skeleton provides a rigid framework over which to add the planking. The type of planking originally associated with skeleton construction is carvel planking, in which the planks are laid edge to edge and are fastened to the frames but not to one another **(Illustrations 9b, 17, 35)**. A V-shaped seam is left between planks to hold the caulking necessary for a watertight seal. Carvel planking is what is seen in movies where bearded, tobacco-spitting sea dogs swap yarns or sing chanteys as they pound oakum into the seams—that is, caulk the seams.

Nautical historians are still debating how and where skeleton construction originated. Some claim it was grounded in techniques that had been used for centuries by Mediterranean shipbuilders. Others maintain it was the invention of Breton shipbuilders who, peripheral as they were to the main Mediterranean and Norse shipbuilding traditions, were able to break forth in a novel direction. A few historians believe it derived from knowledge of Chinese bulkhead construction that had percolated westward. Most experts agree, however, that when it appeared, skeleton construction with carvel planking was essentially a new way to build boats.

Frames vary in size and shape throughout the length of a vessel. A frame amidships in a round-bottomed boat, for example, looks like a broad U in outline, whereas one near the bow is narrower and looks more like a V. The frames must be accurately shaped so that the planking lies flush against all of them and runs fore-and-aft in smooth curves. The main technical problem of skeleton construction boils down to how to obtain accurately the shapes of the numerous frames of a boat or ship that hasn't been built yet. This problem was the bane of early practitioners of skeleton construction, and lack of a good solution caused no end of difficulties. It was the sort of thing that made shipwrights jerk awake at night in a cold sweat after nightmares.

Shipbuilders first using skeleton construction came up against the problem of frame shapes like flies hitting a windshield. Oh, they built ships, all right, but it wasn't easy. The U-shaped frames of wooden ships were massive objects made of hard wood. The dimensions of frames varied with the size of a ship, but as an example, the frames of a ship 70 to 80 feet long might have been 12 inches thick across the face, where the frame contacts the planking, and 16 to 18 inches in depth perpendicular to the face. The wood frequently used for frames was oak, the very wood that provides a common simile for hardness. There were lots of frames, too; a vessel 70 to 80 feet long might have had 40 them. Any imprecision in shipbuilders' ability to pre-determine frame shapes meant considerable trimming of many massive, hard frames.

Shipbuilders literally hacked their way through the problem from the 15th to the 19th centuries. Early on, they used a laborious technique known as building-on-one-rib. A builder (in the collective sense, meaning a master shipbuilder and his shipwrights) first set up the backbone of a ship. He then constructed a single frame amidships and fastened it upright to the keel. He might copy the shape of this frame from an existing vessel he liked, or simply pull the shape out of his head on the basis of the desired beam (width) and cross-sectional shape of the vessel. With this frame in place, he erected a few additional frames fore and aft of amidships, using his experience to approximate their shapes but leaving plenty of extra wood around the edges. He then nailed up a few long, temporary battens, called ribbands, fore-and-aft over the frames, just as the planking would eventually lie. Sighting along the ribbands, he could see where a frame was too wide and caused a bulge. The builder trimmed the edges of the frames until the ribbands lay flush with them in smooth curves fore-and-aft. Once this was done, he could use the ribbands as guides indicating the shapes of the remaining frames to be interpolated between the scattered frames already erected.

By this method, the builder essentially designed the vessel on the slipway as he went along. All of this sounds fairly easy in the writing, until one remembers that the frames being trimmed were massive oaken timbers a foot or more across the face, and that trimming was accomplished with handsaws and adzes. Needless to say, shaping frames in this manner was both backbreaking and time consuming, and shipbuilders strove for ways to reduce the labor and expense.

⌒

A significant advance came in the 18th century, when shipbuilders began carving scale wooden models of a ship's hull before building the ship. They could whittle away to their heart's content, adjusting the shape until the vessel looked just right. Furthermore, because the two halves of a ship are bilaterally symmetrical, the builders needed to model only half of a ship to gain an idea what it was going to look like. Thus arose the hull half-model as a design tool. Half-models were usually carved from a solid block of wood, with the back of the model flat to represent the centerline plane.

Adjusting a hull's design by whittling was much easier than hewing design modifications in oaken timbers during the actual construction. More importantly, the half-model provided a means of pre-determining frame shapes, which could be obtained from the model and scaled up to full size. To do this, the builder fastened the model to a backboard that had vertical lines drawn onto it to indicate the spacing of frames fore and aft. These lines on the backboard assisted him in then drawing the outline of each frame on the model itself as a vertical line running over the surface from sheer (where hull meets deck) to keel.

Various methods existed to lift frame shapes—or, to be exact, the shapes of half-frames—from the half-model. Some builders bent a narrow strip of lead vertically around the outline of a frame on the model surface and carefully removed the strip. The result looked something like half a cookie cutter made from lead. The outline in lead of a half-frame was then traced onto paper and scaled up from there. Other builders used a more direct and accurate approach. With a fine saw, they sawed the half-model nearly through to the backboard at each frame position, inserted a sheet of paper into the saw cut, and traced the outline of a half-frame directly onto the paper.

The half-model was both a design tool—indeed, it was the design—as well as a building tool that allowed the scaling up of frame shapes. It was a vast improvement over building on one rib, but it was still an imprecise method. This is because any small irregularities in the model and errors in lifting frame shapes would be magnified in the process of scaling up.

⌐⌐

Another significant advance was the development of marine drafting techniques that allowed the shape of a hull to be completely represented on paper as a lines drawing, or simply "the lines" **(Illustration 8)**. In this context, "lines" is curiously always used in the plural. A lines drawing is intimidating to the untrained eye. Although the lines do have the pleasing effect of abstract art, with vague boat-like shapes evident among them, the untrained eye will have the devil of a time discerning from the lines what the boat will look like.

A lines drawing consists of three plans, each representing a different view of the hull. There is a front view called the body plan, a side view called the profile plan, and a view from above called the half-breadth plan. The three plans are interrelated through a common reference grid of horizontal and vertical lines. A lines drawing, then, comprises the reference grid and the three plans laid out on a single sheet of paper.

The three plans in a lines drawing are similar to the three views of an object—front, side, and top—typically presented in a mechanical drawing, but there are significant differences. Each view in a mechanical drawing shows the solid surface of an object. In contrast, each plan in a lines drawing shows the outlines of parallel slices, or sections, made at regular intervals through half of the upright hull; as with half-models, it is necessary to represent only half the hull. The body plan shows transverse sections; the profile plan, vertical-longitudinal sections (called buttock lines); and the half-breadth plan, horizontal sections (called waterlines).

Particularly relevant to the centuries-old problem of determining frame shapes is that the body plan within a lines drawing shows the shapes of half-frames at regular intervals along the length of a boat. Furthermore, these shapes are exact to within the thickness of a pencil line at the scale of the drawing. Because the three plans are interrelated through the reference grid, internal checks allow detection of the small irregularities inherent in carved models, especially when the latter are scaled up to full size. A lines drawing is thus a more precise representation of a hull than is a half-model.

One can imagine dramatic ways in which lines drawings emerged from half-models. Perhaps an apprentice shipwright, whose task it was to light the stove in his Master's shop at an ungodly early hour, discovered one cold morning that there was no firewood at hand. There were, however, numerous half-models cluttering the Master's workshop. Reasoning that the models were no longer needed because the ships had already been built, the apprentice used them to start the fire. When the Master arrived to a warm workshop hours later, he found to his horror that his life's work had just gone up in smoke. After flogging the apprentice near to death, the Master devised a system of ordering on paper the information available from half-models, so that this information could be easily copied and stored.

This story would make good light opera, with a countertenor singing the whipping of the apprentice, but it is only fantasy. Marine drafting techniques began to develop in the 16th century, earlier than the earliest known use of half-models. Scale drawings of ships, including series of frame shapes fore-and-aft, have survived from as early as 1585. These were not lines drawings in the modern sense, but they nevertheless contained many of the same elements. In contrast, the earliest surviving half-models are from around 1715.

The two modes of design—drafting and modeling—appear to have originated separately and in parallel. Marine drafting likely developed in a few state-of-the-art shipyards, and the innovators had no incentive to widely disseminate this method. On the contrary, shipbuilding techniques were essentially state secrets, as closely guarded then as they are now. As a case in point, Admiral Frederick Chapman, in gathering designs of boats and ships from all over Europe for his 1768 treatise on naval architecture, *Architectura Navalis Mercaturia*, spent a short stretch in an English prison for possessing what we now refer to as "highly classified material."

Aside from a few thinkers-on-paper working at progressive shipyards, most shipbuilders of the time were unlettered craftsmen. For these men, the development of half-modeling was a godsend, because it provided a tool they understood intuitively and could work with their hands. It wasn't until the 19th century that the parallel traditions of drafting and modeling converged in complete lines drawings coupled with lofting techniques to provide the means of precise skeleton construction as we know it today.

I cannot claim nobody builds models anymore. A marine architect today will fashion a half-model to show a client the appearance of the yacht that will eventually cost the client a small fortune and his marriage. The designer of a high-performance racing yacht might construct a scale model for hydrodynamic testing in a flow tank. These models, however, differ from the design half-models of antiquity in that the architect now draws the lines first and then builds a model according to the lines.

Upscale tourist shops along all the world's coasts sell half-models of yacht and ship hulls, carved in expensive woods, polished to a high gleam, and mounted on varnished backboards. A half-model makes a fine gift for the yachtsperson who has everything else. Few of the recipients of these souvenirs, however, realize that at one time half-models were a primary tool in nautical design, and that the fates of nations depended upon them.

⌒

Today, a marine architect starts out to design a boat with little more than a blank piece of paper and a pencil. It is true he needs a few drafting instruments, but these are only aids in measuring and drawing. His primary instruments are his brain, a pencil, and the paper upon which he draws the lines.

The lines convey the shape of a vessel's hull and provide all the information necessary for a builder to construct the hull exactly to that shape, and the shape of the hull is everything. It is what distinguishes an America's Cup winner from a has-been, a swan from a floating pig. Once the lines are completed, the architect will augment this drawing with many additional drawings showing construction details, the layout of the interior, mechanical details, the sail plan, and so forth. These additional drawings are, of course, very useful to the builder. But even without them, once he has the lines, a competent builder will know how to construct a seaworthy hull, and he will be able to lay out a suitable interior. The lines are the sine qua non of the boat. If you're ever in a position to steal only part of a boat design, make absolutely sure it's the lines.

While a modern lines drawing is intrinsically a more precise starting point than a half-model, the builder of today would nonetheless encounter problems similar to those of his forebears if he attempted to scale up frame shapes directly from the lines, for several reasons. One is that designers occasionally make mistakes in the lines drawing, the table of offsets, or both. The builder eventually catches these mistakes, but if he scales up directly from the lines, he catches them only during building, when corrections are difficult and costly.

Another reason is that, even if a designer were a Leonardo da Vinci in the precision of his lines, inaccuracy is nonetheless inherent in scaling up. Suppose a designer drew the lines of a 75-foot vessel at a scale of 3/8 inch to the foot. At this scale, the thickness of a pencil line—1/64 inch, say—in the lines drawing would mean a margin of error of nearly 1/2 inch in scaling up. This is still enough wood to be irksome to have to trim from the edges of frames erected in position on a backbone.

Finally, the table of offsets derived from a lines drawing provides only a limited number of reference points for a builder to use in scaling up curved frame shapes. The builder would have to interpolate the shapes of the frames between these reference points. Inaccuracies arising from interpolation will again require trimming frames by hand after setting them up.

By drawing the lines full scale on the loft, then, a builder can immediately detect mistakes in the design—but as errors in pencil lines rather than as errors in solid wood. The inaccuracy inherent in scaling up is reduced from the width of a pencil line at the scale of the lines drawing to the width of a pencil line at the scale of the full-sized boat! This is the theoretical level of precision obtainable from lofting. In practice, this level of precision is not possible because other limiting factors come into play, such as the precision with which frame patterns can be lifted from the loft, and frames sawn and erected on the backbone.

The limiting factor in scaling up becomes the skill of the craftsman himself. A boatbuilder who has taken care in lofting, making frames, and setting up the skeleton might have to trim as much as 1/8 inch here and there. If he has to trim 1/4 inch anywhere, he has been sloppy. In any case, even 1/4 inch is a far cry from the ugly amount of hacking builders-on-one-rib had to face in the 15th century.

Chapter 4
More about lofting

In the Bohemian Club of San Francisco there are some crack sailors. I know, because I heard them pass judgment on the Snark during the process of her building. ... She was all right in every particular, they said, except that I'd never be able to run her before it in a stiff wind and sea. "Her lines," they explained enigmatically, "it is the fault of her lines."

—Jack London, *The Cruise of the Snark*

Okay, let's say you want to loft a boat. First you find a tight-seamed wooden floor a little larger than your boat. You whitewash the floor. Then you follow a set order of steps. You start by drawing the reference grid of horizontal and vertical lines. Using measurements from the offsets, you next draw the boat's profile, then the sheer line in the half-breadth plan, then all frames in the body plan, then the curved waterlines in the half-breadth plan, and so forth. You alternate from one plan to another. Whereas the three plans are separate in the lines drawing, on the loft you superimpose them to save space, and the relatively organized curves in the lines drawing come to resemble the uncoordinated scratches of a drunken chicken. Even so, with experience you can easily separate the three plans in your mind.

The most frustrating aspect of lofting is that the points where curved lines cross the reference grid must be mutually consistent among the three plans. Changing a line in one plan requires corresponding changes in the other plans as well. For example, you may draw a nice waterline curve in the half-breadth plan, only to find that this throws one of the frame shapes in the body plan out of whack. So, you go back and forth, adjusting lines ad nauseam until you achieve nice curves in all three plans and correspondence of all reference points.

The process is analogous to working a crossword puzzle. You write a word in Across. You then find that the crossing word in Down gives a different letter at the intersection. You're sure the word in Down is correct, so you look for an alternate word in Across. But the alternate word you choose for Across doesn't have a common intersecting letter with yet another crossing Down word. You adjust words back and forth until the Downs and Acrosses are mutually consistent.

⌐

This is how we beginners spent our first week of boatbuilding school, lofting Trumbly's simple seven-and-a-half-foot dinghy **(Illustration 10).** We drew curves on our lofting boards, checked them, and then erased and re-drew them. With no air circulation, the second-floor loft was as hot as an industrial kitchen. We wore nothing except underwear beneath our carpenter's coveralls but even then poured sweat onto our lofting boards.

When we had lofted the lines of Trumbly's little boat, we found to our consternation that the diabolical exercise had only just begun. We should have seen this coming, for Trumbly had given us his definition of lofting on the first day of class:

"Lofting is the expanding of the designer's lines full size for the purpose of fairing; for measuring frame, stem, transom. and shelf bevels; for making patterns of various kinds; and for correcting possible designer's mistakes. The lines are laid out on a full-sized grid, using the offsets the designer has provided."

What use would it be to loft a boat without knowing how to translate the loft into a real boat? Accordingly, the entire assignment was:

1) Loft the boat.
2) Add a curved, raking (tilted) transom to the lines drawing.
3) Expand the transom (draw the true shape of the transom with the curvature removed) on the loft.
4) Make a master bevel board and a bevel stick.
5) Use these tools to figure bevels for the stem, transom. and frames.
6) Make plywood patterns for the stem, all frames, and the transom.
7) Make the actual stem of the boat from solid wood, including the rabbet groove where the planking intersects the stem.

All this took another few weeks of lecture in the morning and toil in the afternoon. We wanted to get through the assignment fast so we could start work in shop, but when we hurried we made time-consuming mistakes. Furthermore, unlike any previous school we'd ever attended, we weren't being graded on a scale and there was no time limit. Trumbly checked each step, and unless it was up to his high standards we didn't proceed to the next step. Part of the reason we beginners found the lofting exercise tedious is that few of us had built a boat before. We did not know about the backbreaking early years of skeleton construction and thus did not understand a cardinal fact: the one thing more tedious than lofting a boat is to build it without lofting.

Admittedly, another thing more tedious than lofting may be reading about it, and so I will finish this discourse on lofting simply by defining three key concepts that appear in Trumbly's definition and, in fact, permeate all of boatbuilding: fairing, bevels, and patterns.

One purpose of lofting is to allow "fairing." This term stumped us at first; we found it inexplicable, like a Zen concept. In fact, the idea of fairing is simple, and I will it introduce by examining various vernacular meanings of the word "fair." In the eyes of Jesus, to be fair is to do unto others as you would have them do unto you. In sports, to be fair means not fouling your opponent even when the referee isn't looking. In poetry, women described as fair have a pleasing aspect, with the right curves in the right places. The concept of fairness in boatbuilding is most similar to that in poetry.

In boatbuilding, a curved line or surface is said to be fair when it is pleasing to the eye, with no flat spots, humps, or dips disturbing the curvature. As a familiar example, consider the voluptuous fender of an antique Jaguar automobile. If you lay one eye very close to the fender and sight along it, you see a pleasingly curved surface. The surface is fair. However, if some dork knocks a baseball into the fender, the damage may not look too bad from a distance, but if you again sight along the fender at eye level, you will see a slight flat spot or hollow where the ball hit. The curved surface of the fender is no longer fair. It is said to be unfair.

A loftsman must ensure that all the curves he draws are fair, for these curves correspond to the curvature of the hull, and a hull should not have irregularities. To draw a fair curve on the loft, he tacks down a long batten—a straight-edged length of wood thick enough to have some spring to it—to intersect the reference points defining the curve. When the batten is in place, he sights down the length of it, his eyes close to the batten. If the batten runs in a smooth, pleasing curve, then it is fair, and the loftsman can proceed to draw the curve using the batten as a guide. If there are irregularities, he must adjust the batten until it is fair. This process of examining a curved line or surface, and correcting irregularities, is called fairing.

Another purpose of lofting is to measure frame, stem, transom, and shelf bevels. Boatbuilders refer to all sorts of bevels, such as standing bevels, under bevels, real bevels, and apparent bevels, but all these terms are esoterica. "Bevel" means nothing more than "angle." Because boats are full of complex curves, most pieces of wood join other pieces of wood at an angle. A piece cut at an angle across its minimum dimension (e.g., the 1-inch thickness of a 1- by 6-inch board) to meet another piece is said to be beveled, or to have a bevel. A piece cut at a 90-degree angle has no bevel and is said to be square.

Of key relevance to lofting is that the frames of a boat must be beveled on the edge where the planking contacts them. This is easiest to visualize toward the bow. Most frames are positioned transversely in a boat, perpendicular to the centerline, yet the planking runs across their edges at a large angle as it tapers toward the bow. If the frames are not beveled, the planks will contact them only at the forward edge, rather than lie along the whole outer face. Obviously, you cannot adequately fasten a plank to a frame unless the former lies flush across the entire edge of the latter.

The really astounding thing about lofting is that the builder can calculate frame bevels from the loft and cut them directly when he cuts out frames or frame molds, thus again saving considerable amounts of trimming by hand. He can also calculate bevels for other parts such as the stem and transom, and make these parts before setting up the backbone.

Patterns are the ultimate goal and the harvest of lofting; they are the final link between the designer's lines and the actual building of a hull. You make patterns to accurately transfer shapes from the loft to the lumber from which actual pieces of the boat are to be cut. Furthermore, once you have patterns, you can easily build multiple copies of the same hull in production-line fashion.

For the seven-and-a-half-foot dinghy, we made all the patterns required were we going to build the thing. To lift a pattern from a shape on the loft, we tapped in the flat heads of one-inch nails (laid on their side) every few inches along the periphery of the shape. We then laid a piece of 1/4-inch plywood—the usual pattern material—over the nails and pressed down hard so that the nail heads made impressions in the plywood. We then turned over the plywood and used a batten to connect the nail-head impressions with a pencil line to reproduce the shape (this step is identical to the connect-the-dot games you see in newspapers and comic books). Finally, we cut out the pattern out with a band saw and faired the edges with a small hand plane called a block plane.

The nail-head technique is primitive but foolproof. Less than a year after learning it, I got a part-time job in a boatyard, using this simple method to make all the frame patterns for a 54-foot sailboat hull.

᠆᠊

After stressing the importance of lofting, I must now incongruously explain that people constructed wooden boats for nearly 6000 years without lofting them. Great boatbuilding traditions came and went, and some of them were spectacularly successful. The Vikings, for example, built large seagoing boats that carried them from Scandinavia to Iceland, Greenland, and North America in the 10th and 11th centuries, half a millennium before Columbus bumbled his way westward.

The Vikings avoided the problem of pre-determining frame shapes because they didn't use skeleton construction. Instead, they developed to its pinnacle the ancient Norse boatbuilding tradition referred to as "shell construction." In shell construction, a boat is built from the outside inward. That is, the planked outer shell of the hull is constructed first, and only then are the frames fitted inside that. The shape of the hull determines the shapes of the frames, rather than vice versa.

To build a boat, the Vikings laid down the backbone—a keel with a stem attached at each end, for their boats were double-enders. Rather than erect frames on the backbone, they proceeded directly with planking. They began at the keel and laid planks row by row, using the natural curves lumber takes as it bends to obtain the curved shape of the hull fore and aft. To obtain the U-shape of the hull side-to-side, they could vary the angle at which a row of planking diverged from the preceding row.

In shell construction, it is necessary to firmly fasten each row of planking to the preceding row before going on to the next, because there are no frames yet to hold the planks in place. The Vikings fastened planks by overlapping the bottom of each plank over the top of the plank in the previous row, similarly to the way clapboard siding overlaps on a house, or shingles on a roof **(Illustration 11)**. They then drove nails through the overlap of the planks and bent over the protruding ends to keep the nails from pulling out. This method of fastening is called "clenching;" hence, one term for the type of planking the Vikings used is "clenched planking." Since the boatbuilding term for a row of planking is "strake," and the strakes overlap, clenched planking is alternatively called "lapstrake planking." Yet another term is "clinker planking," derived from the Scottish verb "to clink," meaning to clinch or to hold fast.

Shell construction with clinker planking dominated boatbuilding in northwestern Europe and Britain for a thousand years, from the 5th to the 15th centuries. In the 15th century, however, the world came to be viewed as a plum, and larger, faster ocean-going vessels as the means to take a bite of it. As boats evolved into ships, shell construction reached a limit of scale imposed by the method itself. For a plank to be bent with no rigid frames against which to bend it, it must be relatively thin. The planks of the largest Viking boats, for example, were at most an inch thick.

When boatbuilders attempted to scale up clinker-planked boats to the size of ships and to propel them with increasing expanses of sail, they found that thin planking supported by relatively flexible framing was insufficient to withstand the forces of the sea. In boatbuilding terms, the "scantlings," or dimensions, of the lumber that had to be used were too small for the size of the vessels. This limitation of shell construction was reached at hull lengths of around 90 feet.

English kings, of course, exerted their God-given right to exceed any limit they chose. Henry IV launched in 1418 *Grace Dieu*, a massive clinker ship roughly 180 feet long and 50 feet in beam. Built for the Hundred Years War, the vessel was never fitted out due to lack of funds. *Grace Dieu* lay in a mud berth in the Hamble River until it was struck by lightning and burned 21 years after launching.

Seventy-six years later, in 1515, Henry VIII launched *Great Galley*, another monstrous clinker ship similar in length to *Grace Dieu*. Carrying 217 cannons, with sails rigged on four masts and space for 120 rowers, *Great Galley* was the last large clinker vessel built. Eight years after the launching, one critic described it as the most dangerous ship ever sailed by man—meaning dangerous to the crew rather than to the enemy. Henry's experts advised him to tear the thing apart and use the lumber for a carvel-planked vessel.

The 15th and early 16th centuries were a period of schizophrenia in nautical technology. Boatbuilding changed little. Craftsmen continued with whatever local techniques they'd used successfully for centuries to build the thousands of small craft that peppered the coasts of Europe.

Shipbuilding was a different story. In a massive paradigm shift throughout Europe driven by a desire for larger, sturdier ships, the older shipbuilding techniques rapidly gave way to skeleton construction.

Henry VIII was no exception. It is unclear why he ever built *Great Galley*, for in 1510, five years before that clinker-built abortion, he had already launched *Mary Rose*, one of the first British warships built skeleton first and with carvel planking. Columbus's *Santa Maria* of 1492, de Gama's *São Gabriel* of 1497, Magellan's *Vittoria* of 1519—all were built the new way.

We can readily understand the 15th-century shift in nautical technology in terms of a similar shift 500 years later in aeronautical technology. As 20th-century nations sought bigger and faster airships as tools of exploration and national policy, the technology of the propeller gave way to full development of the rocket and the jet. Fabric-on-wood construction gave way to aluminum and titanium. Today we still use propeller-driven craft for recreation and short flights, but use jets and rockets for mass transport, exploration, and war.

〜

At this point, some clarifications will be useful. First, by no means do all boatbuilders loft their boats even today, nor do they all use skeleton construction. Newer boatbuilding technologies did not globally replace older technologies. Boatbuilders use the techniques they have learned and the tools and materials available to them. The Kuna Indians of Panama still hack from whole logs the sailing cayucas they use for near-shore seagoing transportation. Some boatbuilders in Scandinavia still routinely produce clinker boats using shell construction, as their ancestors did a millennium before. In Nova Scotia, boats were being built from half-models as late as the 1960s, and I'd wager good money that someone, somewhere is still using half-models. Alaskan boatbuilder Ed Opheim continued through the 1980s making fine carvel-planked skiffs and dories without lofting them.

Trumbly referred to building a boat without lofting it as "building from the hip pocket," meaning putting a boat together by hook or by crook. He did not denigrate people who built from the hip pocket; he simply knew they could save themselves time and effort by lofting. In any case, it is probably safe to say there are still as many professionals worldwide building wooden boats using idiosyncratic techniques as there are those who loft them.

Another clarification is that virtually any boat, regardless of type of planking, can be lofted. For example, clinker boats, historically built through shell construction, are now routinely lofted. The builder uses patterns taken from the loft to make temporary molds that provide a rigid template for planking prior to bending in frames **(Illustration 11)**. Though the planking is clinker type, the basic mode is skeleton construction.

Finally, wooden boat hulls are built either upside down, or upright. It is easier to plank a hull upside down, but above a certain size it is difficult to flip a completed hull into its proper orientation, due to its large bulk and weight. Thus, skeleton-constructed hulls under roughly 20–25 feet long (depending also on the beam of the hull) are often built upside down on a strongback **(Illustrations 11, 26, 32)**; the frames or frame molds are attached to the strongback by means of cross-members, and the backbone elements are then fastened to the frames. Boats larger than 20–25 feet long are usually built upright **(Illustrations 9b, 16)**; the backbone is erected and braced first, after which the frames or frame molds are attached to it.

Of course, there are exceptions. Ed Opheim routinely constructed skiffs and dories 12 to 25 feet long upright. My classmate Brian Saucier at Bates likewise built a 19-foot, ballasted-keeled, carvel-planked sloop upright **(Illustration 9a)**, and the 19-foot gitney *Devil's Paw* **(Illustration 6)** was likely built upright, due to its wide beam. Conversely, as I describe later, CLK Yacht Crafters built the plug for a 54-foot sailboat upside down **(Illustration 27)**. In these cases, exigencies of construction logically dictated the atypical orientation.

Chapter 5
Unlikely working partners

When one refers to fiberglass boats, what is actually meant is fiberglass-reinforced plastic. ... Plastics have not always been respected. When 1970s Dallas Cowboys star running back Duane Thomas called coach Tom Landry a "plastic man," he presumably was using a vernacular derived from the popular conception of low-cost commercial goods, meaning cheap, fake, and without soul.

—Daniel Spurr, *Heart of Glass*

Patrick Chapman was in his late teens when he started at Bates **(Illustration 12a)**. His mother was a Native Hawaiian and his father a fair-complexioned Navy serviceman. Brown-skinned, round-faced, with curly black hair, Patrick looked much more Hawaiian than Navy. At some point his family had embraced the Baha'i faith, and it was through this connection that Patrick had found the apprenticeship with Ray Tufts, who was also a Baha'i. Unlike most of the students, Patrick had already worked building boats for a couple years when he entered the Bates program.

Patrick strictly adhered to the Baha'i stricture against drinking, and he never swore. His idea of a good time was to sing Peter Paul and Mary songs, accompanying himself on the guitar. Almost reflexively, he functioned as the moral backbone of the boatbuilding class. For example, if someone gouged himself with a chisel and let fly a string of profane or obscene epithets, Patrick would look uneasy and exclaim, "Gosh, it's no big deal—you don't have to swear so much." His great redeeming trait was that he didn't hold debauchery against you; he simply didn't want any part of it.

Curiously, friendships are sometimes based on the attraction of opposites, where the binding forces are mutual respect and the absence of competition. This was certainly the case with Patrick and me. We were 10 years apart in age and a world apart in background. Whereas I was an agnostic and not averse to drinking and swearing, Patrick was deeply religious, with an essentially puritan ethic. All we really had in common were some mutual acquaintances in Kodiak and co-matriculation in the boatbuilding program. Nonetheless, we became close friends and working partners in Tacoma.

After we finished our training in lofting, Patrick and I began work as a two-man team on the 38-foot sailboat under construction. The 38-footer was designated the Trumbly-38, or T-38, because Trumbly had designed it. Trumbly's usual policy was that students worked their first year on whatever class project was underway—in this case the T-38—and then had the option to work their second year on their own boat or small boats commissioned by the public-at-large. But Trumbly was in his last two years of teaching before retirement. The T-38 was far from finished, and he made it clear to us, his ultimate cohort of students, that the big sailboat had priority over any meager personal interests.

"That boat'll go out the door when I do, or we'll all die trying," he said often.

The T-38 was a fin-keeled cruising sloop, with a single mast just forward of center, a long cabin amidships, an aft cockpit, and a curved, reverse-raking transom. The school purchased all the materials for its construction and would sell the completed boat to recoup the considerable outlay, $42,000 by the time it was finished. Trumbly used nothing but the finest materials, and they didn't come cheap.

⌒

Woods are not utilized at random in boats. In the T-38, different woods were selected so that their individual properties matched the functional requirements of different parts of the boat. To begin with, the elements of the boat's backbone—stem, keel, and stern timbers—were of Honduras mahogany. This wood has a reddish-brown color so rich it makes cabinetmakers weep, especially when they think of big chunks of it hidden in the bowels of a boat. Unlike the more familiar red lauan, a soft, tropical cedar often called "Philippine mahogany," Honduras mahogany is strong, fine-grained, rot resistant, and not prone to warping.

The steam-bent frames were of white oak, renowned for its flexibility, strength, rot resistance, and ability to hold fastenings. The deck and cabin beams were also of this wood, laminated into curved shapes. Because of its brittleness and peculiar grain, white oak is unpleasant to work.

The planking was of Port Orford cedar. More fragrant than the Tennessee cedar used for cedar chests, Port Orford cedar gives off a unique rich aroma when it is planed—to me, it seems a combination of cedar, orange, and cinnamon. Trumbly would have nothing less for the planking because the natural resins of Port Orford cedar make it exceptionally rot resistant. It was difficult to obtain, and Trumbly had a standing order for all a small mill in Oregon could supply to him.

The underlayers of the deck and cabin top were cross-laminated strips of western red cedar, a soft, light, easily worked wood nonetheless with good rot resistance. Since the red-cedar layer lay directly over the white-oak beans of the deck and cabin, it was visible from inside the vessel, where it contrasted splendidly with the beams.

The cabin sides were made of one-inch-thick teak. Teak was also used for the outer layer of decking over the red cedar. Because of its oiliness, teak is perhaps the most rot-resistant wood in the world. Left alone, it weathers to a gray surface. Usually, however, boat owners coat it with what resides in it naturally—teak oil—to bring out its rich earth-brown color. Teak's oiliness is also its disadvantage, for few types of glue except epoxies stick to it very well.

Alaska yellow cedar was used for hand railings and ceiling strips inside the cabin. This tough wood gives off a pungent, moldy odor when it is planed. Varnished, it looks like pure gold.

Sitka spruce was used for the boat's hollow mast. A soft, light wood prone to rot, Sitka spruce is also tough and resistant to splitting. It is used for masts because it is stronger for its weight than perhaps any other wood.

Various other woods were scattered around the boat. Mundane local Douglas fir was used for interior framing. Rosewood, rare and very expensive, was used for some of the visible trim inside the cabin because of its red color and intricate grain. Ironbark, a heavy, dark eucalyptus harder than oak and thus even more of a headache to work, was used for the protective guards running along the sheer of the boat. Finally, black-colored lignum vitae—Latin for "wood of life"—was used for shaft bearings. Lignum vitae is the toughest wood in the world, so dense it does not float and so hard that the best way to work it is to turn it on a metal lathe.

This, then, was the royal retinue of woods that went into the T-38. However, this august assemblage of ligneous bluebloods summoned from every corner of the globe included only individuals of the most impeccable character. In other words, only clear, vertical-grain, air-dried lumber was used. A board is clear if it has no knots. It is vertical grain if it has been sawed from the sapwood of a log in such a way that the tree rings run perpendicular, or nearly so, to the face of the board. It is air dried if it has been seasoned slowly at ambient temperature—as opposed to dried rapidly in a kiln, which can cause warping or twisting. Needless to say, these special requirements rocket the cost of already expensive woods right through the roof.

A boatbuilder soon learns that the differences among woods are a lot like those among people. Some woods are white, others black, red, brown, or yellow. Some are soft and easily manipulated, others contrary and cross-grained, fighting any attempt to shape them. Some woods are nearly odorless, whereas others smell of perfume or are as foul as farts. As with people, you never really get to know a wood until you work with it.

When I arrived at Bates, the T-38 was already three years into construction. It had been set up, framed, and partly planked. Topsides, the deck beams were in place but not covered, and the cockpit was framed in. The cabin top had not yet been built, since it is best to leave it off as long as possible for easy access to the interior. Inside the boat, bulkheads were in place, benches and cabinets had been framed, and the engine was installed.

The first job Trumbly assigned Patrick and me was to install Samson posts near the bow. These were two heavy mahogany two-by-eights protruding through the foredeck, about eight inches apart and extending about eight inches above the deck. Samson posts function as cleats for mooring or towing ropes. Named after the biblical hero, they are intended to be strong. Attached to the stem at their lower ends and to deck members near their upper ends, Samson posts are so firmly fastened that any force sufficient to dislodge them will likely take a good part of the bow with them.

When we finished the Samson posts, Trumbly assigned us to lay the red cedar underlayer on the foredeck. Over the deck beams, we glued and nailed a diagonal layer of red cedar strips about four inches wide and a quarter-inch thick. We then glued and nailed on a second layer running at right angles to the first. The strips were thin enough to bend readily over the curvature of the deck. When the glue hardened, the cross-laminated strips formed a strong layer impervious to water. This technique is known as "cold molding."

Cold molding the deck must have been a proving ground. After a week, Trumbly relieved Patrick and me from the task and set two other beginners to it. They didn't fare as well as we had. On their second day, when the glue had hardened on their first day's work, Trumbly spotted gaps between some of the strips.

"Shee-it, I can't leave you little kids alone for a second! This deck has seams you could throw a cat through; it'll leak like a sieve," he shouted over the noise of the shop. He punctuated his critique by shattering the offensive decking with a hammer, pop, pop, pop. "Now get this trash cleared away and do the job right!"

Since we'd passed the cold-molding test, if that's what it was, Trumbly assigned Patrick and me to build the cabin top **(Illustration 12b).** This was much more difficult than our previous jobs, as it required complex joinery. The first step was to fashion the teak members of the cabin log (the base of the cabin) and fasten them to the deck around the rectangular hole left for the cabin. The function of the cabin log was to provide a strong, watertight joint between the sides of the cabin and the deck. After we installed the cabin log, we began work on the cabin sides.

For each side, we scarfed together two one-by-twelve-foot boards of one-inch teak. In general, a scarf joint is used to join two long pieces of wood end-to-end to make a longer piece. To make a scarf, the ends of the two boards are tapered or notched in such a way that, when joined, there is no increase in thickness at the joint. The notched or tapered ends are saturated with glue, overlapped, and firmly wedged or clamped together until the glue has dried. Though no metal fastenings are used, a glued scarf joint is as strong as any other part of the boards, and if done correctly it is almost invisible.

After we competed the scarves, we cut the sides to exact shape as determined by a pattern taken from the loft. We then cut out several rectangular portholes in each side, first drilling the corners with a circle-cutting bit and then sawing out between the holes with a saber saw. We used a router around the circumference of each porthole to provide a lip to bear the glass. When the cabin sides were complete, we shaped the fore and aft end pieces.

Before we could assemble the cabin top, we had to fashion and install four stout corner posts, each rabbeted (grooved) to accept the sides and ends. When all was finally ready, we glued and screwed the cabin sides and ends to the corner posts and cabin log. The last step was to install vertical stiffeners inside the cabin sides. From the top end of each of these stiffeners we ran a long bolt, all the way down through the cabin log, the deck, a deck beam, and the longitudinal carlin below that. These bolts were necessary to firmly attach the cabin to the vessel. The last thing a mariner wants is for his cabin top to tear away in a storm.

Throughout the entire process, we thought long and hard before making any cut. We double and triple checked our measurements. Given Trumbly's exacting standards, the slightest error—one side a quarter inch lower than the other, a porthole half an inch out of place—would have been costly. Even in those days, teak was outrageously expensive; each cabin side was worth a few months' rent for either of us. Despite the pressure, we succeeded flawlessly, and this gave us a dangerous sense of competence and complacency. We had yet to feel Trumbly's lash, metaphorically speaking.

Since we'd done well so far, Trumbly assigned Patrick and me to cover the cabin. We made deck-beam patterns according to a specified curve, or camber, and laminated the beams themselves from oak strips, using iron pegs on a big, iron laminating table to hold the beams in their proper curvature until the glue dried. We then ran the beams edgewise through a jointer to remove the hardened glue that had oozed out. Finally, we cut beveled notches into the cabin sides to accept the beams and glued and screwed the latter into place.

At this point, our beginners' luck ran out.

The next step was to fair the beams, so that the cabin covering would lie exactly flush on them. To do this, we ran a long fairing batten over the beam tops to detect high spots, which we shaved down with a block plane. Fairing should have been straightforward; however, we'd grown overconfident. Tiring of hand planing, one of us had the bright idea to use a hand-held belt sander to speed things up. I no longer remember whose idea it was.

Trumbly spotted us well into the process, one of us handling the fairing batten, the other employing what proved to be a weapon of mass destruction. He ran over, yelling, "Stop, stop, what do you guys think you're doing?"

He was too late. With the power tool, we'd created numerous flat spots in the cabin beams and rounded the edges of many of them so that the cabin covering would not lie absolutely flush. It was like using a butcher knife to do an appendectomy; the patient might live, but there would be some ugly scars.

"Jesus! Little kids! Just when I thought you hotshots knew what you were doing! Well, it's my fault. If the student hasn't learned, the teacher hasn't taught. I should have kept a closer eye on you."

Some of the second-years gathered around and watched as Trumbly silently ran the fairing batten over the beams to survey the damage. They shook their heads almost imperceptibly, as though they were witnessing the aftermath of a bad automobile accident. They didn't say anything, though, or even smirk. When you think about it, it was interesting that Trumbly knew exactly what a belt sander would do to the beams. Either he'd done it once himself, or he'd seen a student or co-worker do it at some point in the dark past.

"This is a good lesson for you," Trumbly said to us. "One thing you'll learn is that no matter how badly you screw up, you can always repair the damage. In fact, almost any idiot can do things right the first time—it takes a real genius to correct serious fuckups."

Seeing the disconcerted looks on our faces, he spread out his arms, palms up, tilted his head back, looked upward as though commiserating with God, and laughed a hearty chuckle. "Ho, ho, ho, little kids!" This was Trumbly's trademark gesture for letting his students know that mistakes were not terminally serious.

Whenever Trumbly conveyed what he considered especially important information, he fastened his wiry hand like a crab claw to your upper arm, putting vise-like pressure on your biceps. Perhaps he feared words were not enough and wanted to establish a direct physical connection, as though he could transfer knowledge at a subliminal, electrical level. Or maybe he only wanted to get your full attention.

Whatever the case, he latched onto my arm—he couldn't reach both Patrick and me at the same time—and proceeded to tell us we'd have to hand-plane the damaged beams square again, glue on oak shims to fill the flat spots, and repeat the fairing process. This we did for the next week. It was astonishing how much damage we'd inflicted in five or ten minutes with the belt sander.

⁓

Trumbly got frequent calls from boatyards needing carpenters, or from individuals wanting their boats repaired, and he would farm these jobs out to students who needed work. One day in October, Trumbly got a call from a guy named Roberts in Gig Harbor who had part-time work for two students. Trumbly sent Patrick and me to see him.

A bearded, pudgy guy in his early forties, Roberts worked as a paint salesman in Seattle. He augmented his income by building small fiberglass boats in his garage. He wasn't a trained boatbuilder. He'd started out first sailing small boats and then built them as a hobby. After acquiring the mold for an 8-foot sailing/rowing dinghy, he'd begun producing boats commercially on a small scale. Until now, he'd laid up the fiberglass hulls and done the finishing woodwork himself. This sort of repetitive production work gets old in a hurry, and he felt things were going well enough to hire out the labor. He wanted us to complete one boat a week, working whenever we had the time.

"It takes me fifteen hours to complete a boat," Roberts said, "but working together, you ought to be able to do one in twelve hours or less. I can pay you $25 apiece per boat."

We pointed out that this amounted to little more than minimum wage, especially considering the time and expense required for us to commute to Gig Harbor.

"Look," he said, "You aren't going to get rich doing this. I'm not getting rich, either. But what you'll gain, in addition to some pocket money, is experience. Let's face it, fiberglass boats are the future, and I'll bet you don't do much fiberglassing at Bates."

Patrick and I said we'd think it over and let him know. At first, I didn't want any part of it. Neither of us had previously done any fiberglassing, and from Roberts's description it sounded unpleasant. Surprisingly, indignant as he was over Roberts's blanket comment on the bright future of plastic boats, Patrick talked me into the job.

"You know," he said, "the guy is right. Even if we don't like fiberglass work, being able to do it might help us get jobs someday. Besides, what else do we have going right now?"

He had a point. So, we arranged to begin with Roberts the following week. He would demonstrate the entire process of laying up and finishing one hull, and then we'd be on our own.

⌒

What is called "fiberglass" when it is hard consists of layers of spun-glass fabric saturated with a plastic. The technical term for this laminate is fiberglass reinforced plastic, or FRP. The plastic component, called "resin," comes as a liquid mixture of two nasty organic chemicals—unsaturated orthophthalic polyester, and styrene. The mixture becomes hard by a process of polymerization, whereby the styrene molecules crosslink the polyester molecules.

Polyester resin will harden spontaneously over a period of weeks if it is left outdoors exposed to UV radiation from the sun. This is too slow if you want to build boat hulls with it. To speed hardening, you add small amounts of an accelerator called cobalt naphthenate and a catalyst called methyl ethyl ketone peroxide, or MEKP. When polyester resin first came into widespread use, someone discovered through unfortunate accident that simultaneous addition of the accelerator and catalyst causes a violent explosion. Therefore manufacturers now sell resin with the cobalt naphthenate already added. All you need to add is the catalyst.

How much catalyst you use determines how fast the resin will harden. If you use way too much catalyst, the resin will "kick"—that is, begin to harden—in the bucket as you stir it, giving off a lot of heat in the process. Cans of over-catalyzed resin have been known to spontaneously combust, leading to boatyard fires.

If you use somewhat too much catalyst, the resin will kick when you are in the middle of a job, making it impossible to roll air bubbles out of the glass fabric. This is bad, for air bubbles result in a weak, spongy laminate. This is the worst case; if your job kicks prematurely, you must wait until the botched layer completely hardens, grind it off, and start again.

If you don't add enough catalyst, the resin will take days to completely harden.

The trick is to add just enough catalyst so that the resin begins to harden soon after you have completed the job. This is touchy, for if you run into an unforeseen difficulty that requires extra time, you sometimes find yourself in a mad race to finish before the resin kicks.

With a mold, it is very simple to build a fiberglass hull. The mold itself is made of fiberglass and resembles a bathtub shaped like a boat on the inside. Using a mold to cast a hull is identical in principle to using a mold to bake an angel cake. You apply the hull in liquid form to the inside of mold, and then let it solidify.

The first step in "laying up" a hull is to wax the mold so that the hull won't stick to it. You then paint or spray on a coat of a special resin called gel coat. Gel coat is opaque white, but you can add color to it if you want a colored hull. The layer of gel coat next to the mold surface becomes the hard, shiny outer surface of the completed hull.

After the gel coat has completely hardened, or cured, you cover the inside of the mold with a layer of mat, a pressed cloth made of glass fibers, something like coarse, white felt in appearance. Using a paint roller, you completely saturate the mat with catalyzed resin. You then use a grooved metal roller, resembling a small version of a paint roller, to remove all air bubbles from the resin-soaked mat.

You put down another layer or two of mat in this manner. When the mat layers are of desired thickness, you apply a layer of heavy, cross-woven spun-glass fabric called roving, again saturating it with catalyzed resin and rolling out air bubbles. The layers of mat give little strength to the hull, but serve as bedding for the rough roving, which provides most of the strength. For larger boats, you use several layers of roving with mat in between them, but for the dinghies we used only a single layer.

The penultimate layer is again mat, and the final layer is a flexible, finely woven fiberglass finishing cloth that leaves a relatively smooth surface inside the hull. After the hull has fully cured overnight, you pop it from the mold, just as you'd flop a baked angel cake out onto the kitchen counter. Presto! You've got an instant boat.

Fiberglassing is indeed an ugly business, but it may not be immediately apparent why this is so. First of all, the resin is a petrochemical mixture with a sickly-sweet, organic odor. The fumes are not good for you, and if you value your health, you wear a respirator mask with charcoal-filter canisters to remove the vapors. The respirator is hot and cumbersome.

The resin is also very sticky. Since it's difficult to work with gloves on, you attempt not to get resin on your hands by using brushes and rollers. At some point, however, only a hand will suffice to push mat or roving into a recalcitrant corner, and you get resin on your hand. Later, you inevitably have to scratch yourself. You do this daintily because your hand is sticky, but nonetheless you deposit resin wherever you scratch.

It is necessary to remove stray resin before it hardens. The solvent for resin is acetone, which is the main ingredient in nail-polish remover. In boatbuilding, you use pure acetone by the gallon to clean your hands, face, hair, tools, clothes, and wherever else you've gotten resin, such as your glasses.

Either respirator masks do not completely absorb acetone fumes, or acetone goes right through skin, because you feel logy the day after a fiberglassing job, like you have a low-level hangover. For days afterward, your breath reeks of acetone. I personally find the odor of acetone to be pleasant, and there are far worse hangovers than the mellow feeling of the acetone hangover. If resin removal were the only down side, a fiberglassing job would be like getting paid to sniff glue with only a minimal threat of collateral brain damage. The liver seems to be rather good at detoxifying acetone.

A more fundamental discomfort is grinding. No matter how hard you try, it is impossible not to leave some rough spots in fiberglassing, and these must be ground down with the same kind of disc grinder used by welders. In addition, if you let one layer harden before you lay on the next layer, which happens when you lay up a big hull, you have to roughen the whole surface with a grinder before proceeding.

The grinder spews powdered resin and tiny glass fibers into the air. The respirator mask keeps this dust out of the lungs. However, no matter how well you cover yourself, the dust manages to penetrate up your sleeves and down your collar, and wherever it reaches, it itches like mosquito bites over poison ivy. In fact, the itching powder sold in gag stores is nothing more than ground fiberglass. Although you can get rid of most of the dust with a shower, some of the tiny glass needles work their way into your pores, causing you to itch for days.

It was no wonder, then, that most of our peers at Bates classified themselves as wood purists and refused to have anything to do with fiberglass. This was an aesthetic rather than a practical choice, however, because Roberts's rosy future for fiberglass was actually the present: most boats under about 60 feet in length for luxury and commercial use (sailboats and power yachts; fishing and charter boats) were already being built of fiberglass. Fortunately, even in fiberglass construction, carpenters are still needed for making patterns; for making the plugs from which molds are constructed; for interior framing, cabinetry, and paneling; and for exterior trim and decks.

The little dinghies were a case in point. After we laid up a hull, Patrick and I spent the bulk of our time adding guards, rails, oarlock mounts, a breast hook, a thwart, and transom braces made from Philippine mahogany or red cedar. The finishing took twice as long as laying up the hull.

Patrick and I got into a routine whereby we made three or four trips to Gig Harbor per boat. Monday or Tuesday after school, we'd clean and wax the mold and roll on the gel coat. On Wednesday or Thursday, we'd spend a long evening laying up the fiberglass, often working until after midnight. On Saturday, we'd arrive early and spend the day installing the trim.

All this involved a lot of travelling and work for $25 dollars apiece. In fact, there was nothing left after buying gasoline and stopping at restaurants after work. We were literally working for experience.

Sometimes extraneous activities associated with the trip to Gig Harbor distracted us from our mission, and our schedule fell apart. One distraction along our route was Eddon Boatyard, where two of our more experienced classmates worked part-time repairing boats. We occasionally stopped to see what they were doing. Our main distractions, however, involved Rob Cramblet, a fellow beginner at Bates. Rob was the tall guy who'd been late for class the first day. He not only looked like a hippie but behaved like one as well. He was something of an anachronism; by the late 1970s, most hippies had turned into mean yuppies.

Rob had recently bought a used houseboat, which he moored near shore along the road to Gig Harbor. To call it a houseboat is generous. It was a leaky, rotten, homemade barge about 10 feet long by 8 feet wide, onto which someone had built a rough shack. It looked like a floating Ozark sharecropper's cabin, and it sank to the shallow bottom if Rob didn't bail it out frequently.

Rob had no vehicle; upon learning we traveled several times a week to Gig Harbor, he began asking for rides. When we arrived at his houseboat the first time, he asked if we wouldn't mind helping him bail it—since we were there anyway.

Bailing for Rob was like playing rugby. The houseboat was top-heavy and unstable. When the barge was half full of water and all three of us inadvertently moved to one side of it, it would start to roll over. In the process of scrambling to the opposite side to stabilize the thing, we'd slip on the rotten, waterlogged wood and go to our knees in the slimy water, banging ourselves up. Bruised, bleeding, and covered with scum after the second or third bailing episode, Patrick finally snapped.

"All right, Cramblet," he said, "that's the last time I'm helping you! This thing is an abortion. You ought to haul it onto land and burn it, or sink it once and for all."

"Aw, c'mon, man," Rob said. "I'm going to fix her up and live on her this winter. I just need to keep her afloat until I have time to work on her."

"You're crazy if you want to live on this deathtrap, but that's your business," Patrick said. "I've got better things to do with my time. "

Patrick came as close to swearing as I ever saw him. I was a little surprised at his vehement refusal. After all, helping your friends was a time-honored Alaskan tradition, no matter how bizarre or protracted were their fantasies. Something other than repeated wet bludgeoning was bothering Patrick. I think what he really objected to was Rob himself, a free spirit given to smoking pot, drinking, and chasing the few wild hippie women that were left. Rob did exactly what he felt like doing on a brilliant fall afternoon, such as pamper a derelict houseboat, and all of this may have deeply offended Patrick's puritan sensibilities.

It wasn't as though Rob had nothing to give in return. He had a 16-foot, gaff-rigged catboat moored in Gig Harbor and took us sailing in return for our help with the derelict, whenever we felt we could interrupt our self-important schedule. The three of us spent some fine, warm fall evenings gliding about Gig Harbor in the catboat.

Nevertheless, Patrick refused to do any more bailing.

We completed at most five dinghies for Roberts, spread over a period of seven weeks due to our distractions. Roberts finally decided he needed to sell some of the boats before making any more. Our job ended in December, which was fine with us. We'd learned all we could from the dinghies, and cold weather made fiberglassing in the unheated garage increasingly unpleasant.

By middle-class standards, Roberts was fairly successful. He had a steady job and a modern, three- or four-bedroom, ranch-style house on a rural road within a half-mile of the seashore. At that time, Gig Harbor was already a pricey area, and his property could only increase in value. He had a wife and two small kids, a hobby that made money, and a garage to do it in.

Maybe because of his modest success, Roberts seemed to look down on Patrick and me in subtle ways. After all, weren't we only attending vocational school? People who see themselves as successful often consider themselves superior to the laborers they hire for things they don't want to do themselves. But it was more than that. Roberts seemed to view us as people of limited aptitude and sincerely wanted to help us. I can't think of a worse form of condescension.

"There's no limit to what a person can do with initiative, planning, and hard work," he lectured us outside his garage on the day we picked up our final wages. "Over time, I'll make enough with these dinghies to put my kids through college. If you guys apply yourselves after boatbuilding school, you could have a house like this someday." He swept an arm toward his domicile as though ushering us into the gates of Heaven.

"But you'll only succeed if you strike out on your own like I have," he concluded. "You'll never get there if you work for someone else your whole lives."

We did the yes-that's-good-advice-thank-you-so-much routine and left. Patrick was unusually quiet as we drove back to Tacoma.

"Jeez, how do you like that guy?" he finally blurted out. "He's got a mortgage, a dumpy wife, a couple of brats, sells paint for a living, and drives a fifty-mile round trip to work every day. And he wants us to be like him when we grow up? He must think we're morons or something!"

That was approximately what had been running through my mind, which was probably why Patrick and I got along so well.

Chapter 6
Daily routine and field trips

Planking is a trade within a trade. It is not that boatbuilders are secretive about planking, but probably because planking is so different from the common branches of woodworking that little has been written about it. Boatbuilders generally are not writers, and the writers are not plankers.

—John Gardner, *The Dory Book*

The first autumn of boatbuilding school ticked by like clockwork, six hours a day, five days a week. It was like being in grade school again— literally. Bates was part of the Tacoma public school system, and we had recess and lunch at the same times as all the school kids in the city. The main difference was that we referred to recess as coffee break.

After roll call at 8 AM, Trumbly would lecture for an hour, sometimes more. After lecture, most of us went to work in the shop. The exception was that students working on design could spend the whole morning in the classroom.

At 10 AM there was a mass exodus to the cafeteria upstairs for coffee and doughnuts. The mid-morning break was supposed to last 15 minutes, but since nobody kept track, we regularly abused this limit. Trumbly himself never missed coffee breaks, as though he considered them a hard-earned privilege of Labor that needed to be defended daily. Instructors had their own small coffee room adjoining the cafeteria.

Starting at noon there was a half-hour lunch break, after which we worked until mid-afternoon break around 2 PM. This wasn't an official break, since the school day ended only an hour later, but usually we dashed upstairs for a ten-minute cup. By this time we realized we hadn't gotten as much done as we'd wanted and were pushing hard.

Around 2:45 PM, Trumbly would shout, "All right, you guys, time to clean up!" from wherever in the shop he happened to be, and the word was passed down the line, "Clean up!" Everyone began brushing off workbenches and returning tools to the toolroom. The task of sweeping the shop rotated among all students on a weekly basis, two students per week. This took a bit of time, for the shop was where we reduced big pieces of wood to smaller ones, using hand saw, band saw, table saw, jointer, planer, hand plane, chisel, drill, slick, and adze. By the end of the day, shavings and sawdust carpeted the floor and lay in drifts under the machines, like ligneous snow. The sweepers gathered these leavings of industry and deposited them in canisters for the janitors to dispose of.

The last soul was out the door not long after 3 PM.

In the context of the daily schedule, I should mention what Trumbly did all day. Obviously, when he was in the classroom, he lectured or helped students with design. When he was in the shop, which was at least half of each day, he spent some of his time on special projects related to the T-38. An example was a gimballed cooking stove that he designed and installed in the galley. In addition, he personally interfaced with the machine shop in having metal fixtures built, such as the mast step, various brackets and braces, and hardware for the rigging.

Sometimes he brought in projects of his own and involved students in working on them. An example was the backbone for the 51-foot sailboat he intended to build at home. He constructed the backbone elements at Bates and used this process as a teaching tool.

Mostly he functioned as a foreman, roving freely around the shop and answering calls for help from students when there was a problem. If we couldn't see him, we had to interrupt what we were doing and search him out, but if he was visible, we hollered out, "Joe! We have a problem here," and he'd hurry to us as soon as he finished whatever else he was doing. He certainly earned his salary, for he was constantly busy, often literally running from one place to another.

Only two types of significant deviation interrupted our usual routine. One was toolroom duty **(Illustration 13a)**. Like sweeping, this was a rotating task that fell to each student for a week at a time, thus approximately three times a year. All Bates-owned hand tools were kept locked in the toolroom and signed out on a daily basis. If someone signed out a tool and didn't return it, he paid for it. The toolroom had a Dutch door where the toolroom guy was supposed to sit to dispense tools.

At first we beginners felt put-upon when it was our turn in the toolroom, for this seemed a gross waste of time. We began to grow suspicious, however, when we observed that none of the old hands complained about toolroom duty. God knows they complained about everything else they didn't like. It was soon evident, too, that the toolroom guy was generally to be found anywhere in the shop *except* in the toolroom. Anyone wanting a tool had to track him down. This seemed to be flagrant dereliction of duty, and we beginners waited for Trumbly to set the errant toolroom guys straight about their responsibilities. But he never did. What we finally learned was that a stint in the toolroom came with an unspoken compensation, which was the time to make our own tools. No one ever said to us, "When it's your turn in the toolroom, you get to make your own tools," but that's how it was.

Some specialized tools used in boatbuilding aren't sold anywhere and therefore must be made by hand. Other tools can be bought, but are much cheaper to make. Trumbly not only encouraged us to make these tools, but he also treated this activity as part of the program—like a classical painter encouraging his apprentices to make their own brushes. The tradition of making tools while on toolroom duty probably evolved spontaneously, and once it had, Trumbly saw it as a way to minimize loss of productivity in other directions. That is, if we knew we'd have roughly three weeks over the course of the year to make tools for ourselves, we wouldn't try to sneak time from other projects—the T-38, for example.

Bevel gauges are an example of tools that are better made than bought. A bevel gauge consists simply of a wooden haft slit down the middle so that a blade can swivel into it. It is similar to a pocketknife, except that the blade has straight sides with dull edges and can emerge from either side of the haft. A bevel gauge is used to record an angle from a beveled piece of wood, or the angle between two intersecting pieces of wood. The actual measurement of the angle can then be read using a separate scale called a bevel stick, or the gauge can be used directly to set the corresponding angle on a band or table saw.

Commercially produced bevel gauges are typically soulless objects made of steel and are too long to fit into tight places. We fabricated bevel gauges of all sizes, especially pocket models 3 or 4 inches long. For these we used scraps of rosewood, teak, or some other fine hardwood for the haft, and shiny brass for the blade. We fastened haft and blade together with a copper rivet though inlaid brass endplates. We polished our gauges like fine jewelry. They were attractive objects, antique in construction, and we vied to see who could make the finest and most elaborate ones. We used these gauges, of course, but made many more than we needed. Those we didn't need, we gave away as presents.

There was a code of etiquette concerning bevel gauges. If someone had just finished what he considered the most eloquent example ever seen on earth, he'd sidle up and silently hand you his masterpiece. It wasn't considered decorous to say, "That's a piece of shit; mine are better," even if you thought so. The only guy who could get away with a negative statement like this was Klarich—more about Klarich later—from whom such an utterance meant he rather liked your gauge. It was considered tasteless, too, to point out obvious flaws like the blade folding too stiffly or not lining up with the haft. On the other hand, it was unwise to over-praise a truly fine gauge, for the owner might interpret this as condescension. No, the proper behavior, whatever the gauge looked like, was to handle the tool, raise your eyebrows, nod your head sagely, and say something innocuous like, "Pretty nice" or "Not bad."

One thing all of us did our first stint in the toolroom was build a long, open, wooden toolbox with a dowel for a handle. We then had a place to store personal tools such as hammer, mallet, chisels of various sizes, sharpening stone, block plane, drill bits, and whatever tools we made. Though all these tools were obtainable from the toolroom, we preferred to use our own, both for convenience and to avoid having to re-sharpen blades someone else had dulled. Furthermore, eventual employment in boatyards would require that we had our own hand tools.

I made many tools over the course of my toolroom rotations. I carved a whetstone holder from solid teak and made a hinged lid for it. I cut the claw off a claw hammer, welded a piece of jointer blade to the head, and ground the whole thing smooth to make a fine adze head. I fashioned the adze's handle from a scrap of laminated oak frame (a commercial yard had donated a stack of unused laminated frames to the boatbuilding program for just such purposes). I made a wooden mallet head from the same laminated stock. On a lathe I turned a length of locust trunk for the head of a caulking mallet **(Illustration 9b)**, and then fitted steel rings to the ends. I made a bevel stick from yellow cedar, marked with lines from which to read angles. When my first toolbox began to overflow with all this stuff, I constructed another toolbox.

In addition to making a lifetime supply of bevel gauges, I made two specialized bevel gauges from patterns Trumbly provided **(Illustration 13b)**. One was a self-reading gauge constructed of brass and teak. The other was a planking gauge consisting of several parts cut from quarter-inch sheet aluminum and then sand blasted to give them a silken surface. The purpose of the planking gauge was to directly read plank bevels from a lined-off hull. Trumbly had patterns for these specialized gauges because he'd invented them, though he never bothered to patent them.

⟿

Another deviation from the day-to-day routine involved field trips **(Illustrations 14, 16, 23, 24, 25)**. We probably averaged one every six to eight weeks, with more in summer and fewer in winter. Usually Trumbly scheduled field trips when he needed to go somewhere anyway. He might get a call from a boatyard asking him for advice, and he'd take the class with him. Or he might require a trip to Seattle for marine supplies and schedule a field trip around that. Even the visits to marine-supply stores served a pedagogical purpose, for it was important for us to learn where to obtain the materials used in boatbuilding.

Trips to Seattle followed a stereotyped pattern. We'd leave Tacoma after roll call, piling into Trumbly's VW bug and students' cars. We'd rendezvous for breakfast at Fisherman's Terminal in Ballard, the main center of supply and repair for the North Pacific fishing fleet. Because so many Scandinavian immigrants—many of whom were fishermen—had settled there over the past century, Ballard was sometimes referred to locally as "Little Norway." Trumbly always ordered buckwheat pancakes, and out of curiosity many of the students ordered the same. After breakfast we'd tag along as Trumbly did his shopping for bronze fastenings, marine paint, or whatever, and finally we'd do the field trip.

A memorable trip was to Coolidge Propeller, a foundry that specialized in bronze propellers **(Illustration 14)**. The foundry was of great interest because Trumbly had recently taught us how to design propellers and make the wooden models necessary for them to be sand-cast **(Illustration 15)**. Trumbly was nearly as fanatical about propellers as he was about lofting. In the 1950s and 1960s, he'd designed, built, and raced D-Utility class powerboats—minimal speedboats designed to carry a person, an outboard motor, and a gas tank. You could call it a hobby, although Trumbly was one of those rare people whose profession and hobby are indistinguishable. Anyway, in his powerboat phase he'd learned propeller design to increase the performance of his racing boats. He got good enough at it that he sold one of his propeller designs to Budweiser for exclusive use on their promotional racing boats.

The foundry, of course, had much bigger fish to fry than small sailboat and powerboat propellers. Their bread and butter were massive propeller blades for ocean-going ships, each blade as big as a sea lion and weighing tons. The foundry also produced propellers for fishing vessels, but only in large orders. I suspect that if any two-bit yachtsman had walked in off the street asking to have a single small propeller cast, they'd have quaked with laughter. But Trumbly wasn't any two-bit yachtsman. They knew him at Coolidge, and it didn't hurt that one of the foremen was a former student.

As it was wherever we went, Trumbly greeted the workers, "How are you, so-and-so, good to see you again!"

He introduced the class to his former student, a middle-aged man stout as an anvil from heavy work, with hands like baseball mitts, scarred from metalwork, "This is so-and-so, I knew him years ago when he was just a little kid, like you guys."

And he bantered with the guy, a "You remember that propeller you guys cast for me for the Sacramento River race? Damn thing fell off; nearly ruined my racing career! It was the best propeller I ever had; I almost dove overboard for it."

Trumbly was in his element on field trips, revered among his peers. Bates was the Harvard of boatbuilding programs in the United States, and Trumbly was a long-tenured professor. He played his role to the hilt and reveled in it. Then again, it was rare to see him other than in his element. He seemed at ease whatever he did, with whomever he spoke, and wherever he was. He was at ease with himself.

Seeing Trumbly hobnobbing with workers in the yards, you'd certainly not have suspected that he himself knitted the woolen mittens and watch caps he wore in cold weather. When he first told us this, we didn't believe him. The next day, he brought his knitting to class and gave a demonstration. He was good at it.

"A person ought to be able to make or fix anything he uses," he said.

⌒

The only time I saw Trumbly slightly out of his element was during a field trip to Evergreen College in Olympia, Washington. Evergreen was smirkingly known as an "experimental" or "innovative" institution, where students had great leeway in planning idiosyncratic courses of study and getting credit for almost anything they did, on campus or off. We joked that you could find courses there like underwater oil painting and llama shearing. The traditional A to F grades were replaced by a pat on the back or encouragement to do better. In this last regard, Evergreen was much like Bates.

Some Evergreenies had decided they wanted to study wooden boatbuilding. We Bates students considered this a hilarious prospect— to think that the longhaired, dreadlocked, tie-dyed, pot-smoking progeny of the beatnik generation could build anything more complicated than a prefabricated yurt! No doubt, we remarked disparagingly, the Evergreenies thought they were going to sail the boat around the world, studying oceanography and cultural anthropology along the way. Ironically, this was probably exactly what they intended to do with it.

Trumbly didn't share our prejudices against the Evergreenies. He'd gotten a call asking him to help them line off their framed hull for planking, and this was enough for him. The Evergreenies could have been ax murderers, for all he cared. They were building a boat, and this made them members of the same race, religion, and creed as himself.

One brisk winter day we all piled into automobiles and drove to the college. We Bates students had expected to see something like a large rowboat cobbled together with scrap lumber, bent nails protruding. What we were shocked to see was a full-keeled, round-bottomed cruising sailboat even larger than the T-38 **(Illustration 16)**. Bent-oak frames and deck beams were in place, and the hull was indeed ready for planking. Unexpectedly, the craftsmanship was sound; whoever directed the project knew what he or she was doing. This was a tremendous pinprick to our over-inflated sense of superiority. Not only were the Evergreenies academically better rounded than we were, but they appeared reasonably competent in what we considered our domain. Furthermore, they surely had more fun than we did, because some of the Evergreen boatbuilders were women.

⤸

When a layman thinks of planking (admittedly, the topic does not often come up at laundromats in Iowa), he generally thinks of carvel planking, which is the most common type on larger wooden vessels. When I started at Bates, it was my misconception that the carvel planks on a boat are all of uniform width, like house siding. Although this may be true of slab-sided scows, it is not the case for round-bottomed vessels. Like a pregnant guppy, a round-bottomed boat is distinctly fuller in the middle than at the stem or stern. Therefore, the planks must be wider amidships than near the ends of the vessel. A familiar analogy is the strips of skin sewn together to make a football; these are wide in the middle of the ball and taper to the ends. Unlike a football, however, the planks of a boat are not even of uniform width at a particular place along the hull, but instead vary in width between the sheer and the keel. They are typically wider near the sheer, for example, than at the extreme curve of the bilge, where it is impractical to fasten a wide plank to a sharply curved frame. The garboard plank, the bottommost plank lying just above the keel, is generally wider than any of the rest and is triangular **(Illustration 17)**.

"Lining off"—the reason for our trip to Evergreen—is the first step in planking a boat **(Illustration 16).** The goal of lining off is to mark the edges of every plank onto the frames. This is done before any of the planks themselves are cut, and indeed is necessary to determine the shapes of the planks. Lining off ensures that the planks will be of appropriate widths and run fore-and-aft in a manner pleasing to the eye. In other words, it prevents the builder from getting halfway through actual planking before realizing the job is going to look like crap and wishing he'd done it differently.

To line off a hull, a long batten is first used to "strike"—that is, draw—four fair lines the length of the hull on one side: near the sheer; above the turn of the bilge; below the turn of the bilge; and at the top of the garboard plank. Only one side need be lined off because theoretically planks on the other side of the vessel will have identical outlines. These four lines divide the hull into three longitudinal sectors running from stem to stern, like the sectors of a banana peel. The builder then decides how many rows, or strakes, of planking he wants in each sector, based on the approximate desired width of the planks in each sector amidships at the widest part of the hull. For example, he might decide on 8 strakes in the top sector between the sheer and the bilge, 11 strakes around the turn of the bilge, and 6 strakes between there and the garboard.

The builder then makes a planking scale called a "Boston batten" for each of the three sectors. This is a kind of homemade ruler—a scrap of wood long enough to span a sector at its widest point, with units marked onto it according to some devious Yankee magic. A Boston batten allows the builder to read directly the necessary widths of planks in a sector on any frame fore and aft. For example, suppose the builder has decided on seven strakes of planking in the top sector. Using a Boston batten, he determines the planks are to be 5 and 1/8 inches wide on a frame amidships. The builder moves three frames forward, holds the Boston batten up to the width of the sector there, and directly reads that each plank must be 4 and 7/8 inches wide at that point—the planks are already tapering on their way to the bow.

In practice, the builder determines plank widths, and measures and marks the edges of the planks, only on every third frame. When all plank boundaries have been marked in this fashion, he then tacks up a fairing batten along the marks for each plank boundary and draws the plank line on all frames.

Another aspect of lining off involves determining where the ends of the planks will abut one another. Clear, vertical-grain lumber of planking quality generally comes in lengths of 12 to16 feet, so no one plank can extend from stem to stern on a longer vessel. Planks in a strake are cut so as to abut one another between frames, rather than on them. To join plank ends, a short length of wood called a "butt block" is glued and screwed to both planks inside the vessel, spanning the seam. The resulting joint is called a "butt joint," or "butt."

Butts are weak spots in the hull. To prevent a local concentration of weak spots, there is a rule that butts must be separated by at least three strakes, or three frames, or any combination thereof. For example, butts can be two strakes and a frame apart, or two frames and a strake apart. It's a rule similar to that in chess for moving a knight.

Boatbuilders use a "butt chart" to keep track of butt positions. This is simply a piece of plywood on which a large horizontal rectangle has been lined off into rows and columns, like an accountant's ledger. Each column corresponds to the interval between two frames, and each row corresponds to a strake. Before any planks are cut, the builder marks all butt positions on the chart. With few exceptions, butts are in the same positions on the port and starboard sides of the vessel.

Trumbly gathered the Evergreenies alongside their boat. In the cold winter air, he delivered an impromptu lecture on the principles and practice of lining off. He then divided them up into teams, each including a few Bates students, and directed the work like a conductor. When the first batten had been tacked to the frames, Trumbly stood atop a sawhorse twenty feet in front of the boat and demonstrated how to sight down the length of the batten to fair it. Ever the showman, in his white coverall and baseball cap, arms out for balance, he looked like an egret about to launch into the air.

Several Evergreen faculty members were involved in the boatbuilding project. One of them was an English professor who required the students to keep a journal on their feelings as they built the boat. This was an encumbrance we fortunately escaped at Bates, where feelings were far down on anyone's list of priorities. Also involved was an architect, a visiting woman professor who'd been a disciple of Frank Lloyd Wright. At that time, I didn't know Frank Lloyd Wright from Ho Chi Minh, but I gathered he was someone important from the reverent way the Evergreenies spoke about the woman.

"Oh yes," they said, "she's very good—she studied with Frank Lloyd Wright."

Only much later did I learn that Wright had spawned hundreds of disciples at Taliesins he founded at all points of the compass.

Shortish and heavyset, the woman wore a richly patterned muumuu with a string of costume pearls. She bustled about, framing the boat or parts thereof with her hands, casting architectural terms port and starboard, waxing eloquent.

"The boat shape is so natural, don't you think? The form of the hull is certainly the environment with which the cabin must be integrated." Or some such rot. To give the woman credit, as a participant she had to say something, even though she was out of her element.

Whether Trumbly had planned on it or not, he'd been scheduled for a luncheon with the Evergreen faculty participating in the boat project. The ostensible purpose was to discuss the design of the boat's cabin. None of us Bates students was invited; those of us who'd brought a lunch ate it, and the rest went hungry. We wandered around the campus to keep warm.

When the luncheon was over, it was early afternoon and time to return to Tacoma. Trumbly was uncharacteristically quiet as we walked back to the parking lot. Finally Patrick, whose idle curiosity had no limits, asked, "Well, Joe, what went on at your meeting? Did you decide on the cabin?"

Trumbly shook his head, a pained look on his face. "That woman may be a fine architect, but she doesn't know anything about boats or sailing. She'd have the Taj Mahal sitting on the deck, if she could." Coming from Trumbly, these were damning comments indeed. As an afterthought he added, almost to himself, "You can't design a boat by committee. It just doesn't work, in my opinion."

⤝

The whole trip to Evergreen seemed on the ill-starred side of the astrological scale, a little out of kilter. On the way back to Tacoma, some of us rode with a student named Willie Hartman in his new-model Dodge Charger. It wasn't long before Willie and another student, Greg, began playing road tag, passing one another and speeding ahead. This had gone on for some miles when Willie noticed a state trooper's flashing lights in his rearview mirror.

Willie pulled over, and the trooper stopped behind and got out of his car.

"What's the problem, officer?" Willie said.

"The problem is that you've been doing eighty in a fifty-five mile-an-hour zone, and you've been racing with your buddy in that tan Toyota up ahead."

"Tan Toyota? I'm sorry, sir, but I don't know anyone with a Toyota."

"Uh-huh," the trooper said, gazing at Willie as though Willie were a hybrid between a garden slug and a used condom. Then the trooper pointed his right index finger up, his hand held shoulder high like you see on plastic statues of Jesus, and glanced toward the sky.

Willie craned out the window so he could look up, and the rest of us did the same out other windows. A traffic surveillance airplane circled far above the car. "Shit," said Willie.

"Yep," said the trooper.

Willie was a big man in his early twenties, a bona-fide black-haired Alaskan Native of Aleut-Koniag-Russian descent. I came to know Willie better than I knew any of the other second-year students because he also was from Kodiak, though I hadn't known him there.

Willie came from a commercial fishing family and from an early age had seined salmon in the summers. Fishing kids grow up fast. Around Kodiak I saw 12-year-old Native kids driving mammoth seine skiffs, working 18-hour days in one of the world's most grueling occupations. I imagine this was how it was for Willie. By the time he graduated from high school, boats were as familiar to him as a bicycle is to the average teenager. Boatbuilding was a natural extension of what he'd been doing since childhood.

Fishing kids also grow up exceptionally strong, due to the manual labor involved in fishing, perhaps combined with good nutrition and a genetic predisposition. That was Willie, barrel-chested and strong as a bull sea lion. Centuries ago, at the time the Russians invaded Alaska, he would have been a "strongman." In those days, the Aleuts would recognize young boys of his physique and train them throughout childhood for great strength and stamina. There are stories of strongmen who could lift a fully loaded kayak and its occupant completely out of the water and over their heads. The strongmen were the Aleut equivalent of cruise missiles, which the elders of the tribe would unleash on enemies when diplomacy failed. Strongmen lived short lives because of the hardship of their training and were accorded great honor for this sacrifice.

The cruise missile I knew as Willie was a gentle, quiet, good-natured man who would have been horrified to hurt anyone. He studied at Bates during the day and worked at the Martinac shipyard after school. He was still a fisherman at heart, though, and partied hard on weekends.

~

Lining off the Evergreen boat was fortuitous for those of us in our first year at Bates. Although Trumbly lectured on the topic, we would not otherwise have had an opportunity to observe it in practice. The T-38 had been lined off the previous year, and the planking was half completed.

My first fall at Bates, Trumbly had two two-man crews working to complete the planking on the T-38. Other students envied them, for although there are many jobs in boatbuilding as difficult as planking, few of them have the mystique attached to it. Planking is a contradiction. It is very heavy work, but it is also an exact art. It has to be exact because the planking is what keeps the occupants of a vessel from drowning.

It finally dawned on me that every single plank in the T-38 was unique; there was not another exactly like it anywhere else on the vessel **(Illustrations 9b, 17, 35)**. Planks near the bow and stern tapered markedly toward one end. Those at the turn of the bilge were hollowed on the inside surface to allow for the curvature of the frames. Planks lying in the tight curve where the frames turned toward the keel were convex on the inside. The garboard plank was triangular, while the planks just above it were narrower in the middle and had curiously widened ends. Finally, though every plank on one side of the vessel had a "sister plank" of identical outline on the opposite side, the two sister planks were not truly identical but, due to the bevel along one edge, were mirror images of one another.

Though the planking crews did most of the planking, Trumbly required every student to shape and install a pair of sister planks, to learn how to do it. There were enough to go around. The T-38 had 25 strakes, or rows of planking, on each side. Assuming an average of 2.5 planks per strake, this gives roughly 62 planks per side, or 124 planks for the whole vessel.

Once a framed hull has been lined off, there are two ways of determining the shape of each plank to be cut. Trumbly had each student use both methods in order to learn them, which was possible because we each made two planks. One way, an old method called "spiling," involves tacking a long batten where the plank will go, taking width measurements at intervals from the lines on the frames, and transferring these measurements to a piece of planking stock. The other way, called the "direct method," involves tacking down a long, narrow batten along each of the two lines delineating a plank on the lined-off hull, laying a long strip of plywood over the battens, and tracing the plank shape directly onto the plywood to make a plank pattern.

For a one-off hull like the T-38—that is, a design for which only a single hull is to be built—it might seem a waste of effort to make a plywood pattern for every pair of planks. After all, each pattern will be used only twice, and will then have no further use. Spiling might thus seem to save time and materials. Nonetheless, Trumbly insisted we make a direct pattern for every pair of sister planks and keep it until the planking was finished. He claimed patterns are more accurate and save time over spiling even if they are only used twice. Furthermore, they save time if a plank breaks during installation, which occasionally happens.

Trumbly wanted us to make plank patterns because this had been standard practice in production wooden boatbuilding, and he wanted to teach us that tradition. His generation was the last to build wooden boats on an assembly-line scale. After WWII, for example, Trumbly and his brother Fritz working at Puget Sound Boatbuilding helped build more than 50 bent-framed, carvel-planked, 30-foot gillnetters in eight months. This averaged to roughly one finished boat every 3.5 days, assuming a five-day workweek. Patterns made this speed possible. Plank patterns for the gillnetters were each used not only twice, but 100 times. Patterns allowed all the planks for a boat—indeed, for many boats—to be cut before a single plank was fastened. There were crews devoted to cutting planks using a machine called a tilting-arbor plank cutter **(Illustration 18)**, and they became very fast at it.

Chapter 7
Sketches of some students

Dinghies are a bit like people: they come in many shapes and sizes, but it's unlikely that a single one of them is perfect.

—Stan Grayson, *The Dinghy Book*

After Trumbly trained us beginners in lofting, he began to teach us design **(Illustration 1)**. This sequence made practical sense. Through lofting, we had learned the anatomy of boat lines in excruciating detail. This is some of the knowledge necessary to create the lines in the first place.

As an introduction to design, Trumbly required every student first to draw the lines for a V-bottomed planing hull. After that, we progressed to a modern, 40-foot, fin-keeled sailboat; in addition to drawing the lines, he required that we make all the necessary stability calculations and draw the sail plan. This exercise continued well into the fall, during which Trumbly lectured on design for the first hour every morning. Though the second-year students had already been through the design sequence, they got a second dose of it, because Trumbly strongly believed in repetition as a teaching tool. After morning break, we beginners would return to the classroom to work on our designs while the rest of the class went to work in the shop. Afternoons, we were in the shop like everyone else.

An unpleasant incident occurred during one of the lectures on design that illustrated Trumbly's no-nonsense, nuts-and-bolts approach to training—and that Bates was not for everyone. During the fall, several second-year students graduated from the program, opening slots for newcomers. One of the newcomers was a guy named Bob, who had finished college the previous spring and arrived at Bates with an attitude: he felt superior to the rest of us by virtue of his college degree.

I didn't know how many of my fellow students had attended college. Tom Mankin, the retired Air Force officer, was almost certainly a college graduate, though he never mentioned the fact. Similarly, I told no one I was a college graduate. It was not only irrelevant information, but I was also a little embarrassed. I thought I'd have a bit of an edge because of college, but soon learned that I was as much of a klutz in the shop as the other beginners. And why shouldn't I have been? Dissecting a fetal pig or synthesizing aspirin had no bearing whatsoever on fairing a cabin top.

Bob, however, felt differently. He let it be widely known that he'd recently graduated from college with a degree in geology. He sat in lecture every morning with body language that cried out, "I really don't belong here; this is very primitive stuff you're all doing, but I tolerate it because I'm a student of life." To hammer home his message, he wore a tweed jacket with leather elbow patches.

Bob almost hadn't made it through the lofting exercise, which he declaimed as tedious busywork. Nonetheless, we tolerated him, mainly for his entertainment value. It was interesting to speculate how his attitude would play out. Indeed, he might eventually have adjusted to Bates had not his condescending demeanor also extended to Trumbly.

It all came to a head one day when Trumbly was lecturing on Simpson's formula, which is used to estimate the volume of a boat hull and thence the hull displacement. With a polar planimeter, you measure the cross-sectional area of the hull at an odd number of stations in the body plan of the lines drawing, and then simply plug these numbers into the formula. After presenting an example, Trumbly asked, "Are there any questions?"

Bob's hand shot up. "You know," he said, "wouldn't this be a lot more accurate if you used calculus?" Bob had clearly learned something in college, for he knew how to phrase a statement as a rhetorical question.

"Yes, I suppose so," said Trumbly, rubbing his chin as though this were something he'd never considered before. "Tell me, Bob, have you studied calculus?"

"Yeah, I took it a couple of semesters in college," said Bob proudly.

At this point, I could see what was coming, for I also had studied enough calculus to know that it would be difficult to apply to sailboat hulls. Furthermore, even if it had been feasible, I couldn't have used calculus because I'd forgotten it all.

"That's great that you took calculus!" said Trumbly. "If you wouldn't mind, I'd like you to come up here and show the class how to calculate the volume of this hull using calculus." He held out a piece of chalk to Bob.

There was a long pause, during which Bob was undoubtedly counting the massive proliferation of flies in his ointment. "Well," he said, "I don't think I can do it, but I do know that calculus would be a better way."

"I gotta tell you," said Trumbly curtly, "you're not the first guy who's suggested I use calculus instead of Simpson's formula. But, it's damn strange that every time I ask someone who's studied calculus to come up and apply it, he can't. So, tell me, Bob, what's the good of knowing calculus if you can't use it?"

The classroom was silent. Bob flushed from chagrin, red as a whore's lamp. His mouth worked like a suffocating fish's, trying to produce an answer that wasn't there. Trumbly, having tolerated Bob's disdain for a month, lost his brakes and rumbled out of control down a rocky slope.

"The fact is, to use calculus you need to know formulas for the lines of the hull, and we have no way to determine these formulas. And even if we could determine them, boatbuilders who know calculus are as rare as two-headed cows. This is why we use Simpson's formula. It's simple enough that anyone who knows basic arithmetic can use it. It's a rule-of-thumb method, but the answers it gives are accurate enough. Next time you visit a marina, keep in mind that every sailboat you see there was designed using Simpson's formula."

I never understood why Bob entered the boatbuilding program. Perhaps he'd applied to Bates in a moment of panic and depression during his last years of college, or maybe he'd thought boatbuilding would be a noble, pipe-puffing sort of hobby for an educated gentleman. In any case, Bob quit Bates not long after the calculus incident, vacating his slot for another person.

⌒

In a kind of karmic balancing reaction, this person turned out to be Klarich, a guy about my age who proved to have an almost preternatural talent for boatbuilding. Klarich was the direct opposite of Bob and would have been as shaky in a college environment as Bob was at Bates. Klarich did have a first name, which was Mark. In those days I called him by both names— either Mark or Klarich—but I now think of him as just Klarich. I've met several Marks in my life, but only one Klarich.

Mark Klarich **(Illustrations 14, 37a, 39)** was an iconoclast. He drove a white, sixties-model Ford pickup that he called White Cloud. He wore his blond, Slavic hair in a shoulder-length pageboy style and was constantly brushing it out of his eyes. He was about my height, a couple inches under six feet, and of average build, but he was strong and wiry. This physical toughness, combined with his forceful personality, made him seem larger than he was. Klarich was one of only two cigarette smokers in the class. On breaks, he often carried his coffee down to a sunny spot in the courtyard and warmed himself like a lazy iguana, an unfiltered Camel in his hand. In the cool winter months, he wore a shaggy, oversized woolen sweater that made him look like an unfiltered camel himself.

Klarich had an ugly scar across one side of his forehead. Through his twenties he'd worked various jobs on the Flats, where he'd done a lot of drinking in the working bars. He got the scar late one night when he exited a bar, well lubricated, and stood alongside the street waiting to cross. Someone opened the door of a pickup truck speeding by and caught him in the head. Given Klarich's sometimes-abrasive personality, it's possible that this was intentional. Whatever the case, he woke up in the hospital with a concussion, and after his head healed externally, he was subject to random epileptic fits.

Klarich's father was a butcher. Although Mark helped out on Saturdays at his father's shop in Tacoma, his heart wasn't in the meat business. I never asked why he chose to pursue boatbuilding rather than the family trade, but maybe he considered boatbuilding his birthright. Yugoslavian immigrants from the Dalmatian coast had laid the foundation of the boatbuilding industry in the Pacific Northwest: Barbare, Martinolich, Martinac, Petrich, to name a few. Klarich wasn't from one of the boatbuilding families, but he surely had some ancient wood-hewing gene in his blood.

Klarich and I became friends, ironically over an episode that might easily have made us enemies. After Patrick Chapman and I finished the cabin on the T-38, Trumbly moved Patrick to some other task and assigned me to install glass in the portholes we'd cut in the cabin sides. Compared to the cabin itself, this seemed as trivial as a drunken afterthought. It was a simple matter of fashioning four pieces of teak molding for each port to hold in the glass. To obtain the shapes for the molding, I held up a piece of plywood and directly traced out the outline of each port. I soon made molding for all the ports. The patterns for the molding also served as patterns for the glass. It was some sort of expensive leaded glass that had to be specially cut at a shop in South Tacoma.

The glass should have been easy to install. All I had to do was insert it in the rabbeted porthole, bed the molding around the glass using a sticky, anti-rot compound that we called "bear shit" to give a watertight seal, and fasten the molding in place with small screws. However, when I installed the first port, the glass cracked. Thinking this had been an unfortunate fluke, I installed the rest of the ports, and they also cracked. Those that didn't crack immediately were cracked when I came back the next day, as though an evil genie had tapped them in the night with a ballpeen hammer.

One technical problem was that the portholes had been cut and rabbeted with the cabin sides flat. Installed, the cabin sides were slightly curved, and thus the rabbet flanges of the portholes were also slightly curved. But I'd anticipated this, and had spent a lot of time with a chisel making the flanges flat, fitting in the plywood patterns to make sure.

After the first round of glass cracked, I spent more time making sure the rabbets were flat. Then, rather than crack all the pieces again, I decided to work on one port until I could get it safely installed. Only then would I do the rest. I worked on this port for days, cracking it in the morning, going to the glass shop in the afternoon, and trying again unsuccessfully the next morning. Once I was able to crack it twice in the same day. This was not just expensive; it was insanity. Wasn't it Einstein who defined insanity as repeating the same experiment multiple times with the hope that the outcome will change?

One morning Klarich, who I hardly knew at the time, approached me while I was in the process of removing yet another piece of cracked glass and said, "You're kind of a slow learner, aren't you? Don't you know what you're doing wrong? I'll bet I could do all those ports in a day, without cracking any of them."

That was Klarich's way of speaking, openly and frankly. You didn't know whether he was close to Buddhahood, or was just yanking your chain to see what would happen, or maybe both. Whichever the case, I was ugly with frustration and wanted to strangle him.

Instead, I said, "You want to try the sonofabitches? They're yours!"

We notified Trumbly that Klarich was going to have a go at the glass installation. Trumbly readily assented. Although he hadn't said anything to me, he must have been growing concerned at the rapidly mounting expenditures for leaded glass, as well as the wasted time.

As I worked on another project, I kept a close watch on Klarich out the corner of my eye, expecting him to humiliate himself. After all, he was the rankest of beginners, several months junior to me in the program. But damned if he didn't get all the ports installed in short order, and they stayed uncracked. My only consolation was that it took him two days rather than one. When his success was sure, Klarich attempted to explain to me the right way to do the job, but I ignored him. He was too smug by half, and I was sick of the whole thing. I'm embarrassed to say that to this day, I don't know what I was doing wrong.

The only predictable thing about Klarich was that he was unpredictable. For example, he had a big, healthy, pretty, amiable girlfriend of Norwegian descent named Kathy who was as steady as Klarich was erratic. She was studying toward a horticulture degree at a local community college, intending eventually to start her own landscaping or greenhouse business. I thought they made a nice couple, but once when I asked about her, Klarich said, "She's doing fine. But, you know, I'm kind of tired of her. Do you want her? I'll give her to you."

I reacted just as he probably hoped I would. "Jesus Christ, Klarich! What wrong with you? You don't own her; you can't just give her to someone."

"Well," he said, grinning like the Cheshire Cat, "she is mine right now, but if you want her, you can have her."

Looking back, I have to admit I probably did want her, but it wouldn't have worked. If I'd gone after her, Klarich would suddenly have decided he wanted her back, and that could've gotten ugly.

Klarich told me he lived in a chicken coop, along with two cats. When I finally saw his place, it proved to be a large chicken coop that had been remodeled by the landlord into a small, multi-roomed bungalow complete with running water and electricity. It was hard to tell it had been a chicken coop. Of the cats, one was normal, but the other was crazy. It might have had agoraphobia. When Klarich put the insane cat outdoors, it huddled up and shivered, with a grimace on its little cat face, as if predation were both predestined and imminent.

"I love this cat," Klarich told me, "because I understand how he feels. Look at him there, staring up at the sky, whimpering. My life is just like his. I whimper, uuuuuh, uuuuuh, uuuuuh, and wait for the hand of God to fall."

⌐

Just as in lofting, a specified sequence of steps is followed in design. However, design is a much less mechanical process than lofting, for the designer must have some idea how the shape of a boat hull—in this case, a sailboat hull—relates to performance in the water. It was thus a slight problem was that none of us beginners was an experienced sailor, and Patrick Chapman and I had scarcely set foot on a sailboat. The assignment was like asking a casual airline passenger to design an airplane.

Seeing that this was a problem, Trumbly did what he could to remedy it. The opportunity arose one day during the fall when he needed to move *Osage*, his 40-foot wooden ketch, from a mooring near his home on Raft Island to a berth in the Tacoma Yacht Club. Trumbly had designed *Osage* himself, built her at home over a period of seven years, and sailed her casually for five years. However, like many boatbuilders, he was fundamentally more interested in building boats than in using them. He'd recently sold *Osage* so he could buy materials for his next project, a 51-foot, cold-molded sloop he intended to call *Windance*. The reason for moving *Osage* was to deliver her to the new owner. Trumbly asked Patrick and me if we'd like to go along, ostensibly to help, though he didn't really need any help because he'd rigged *Osage* to be sailed single-handed.

One windy fall Saturday perfect for sailing, the three of us set out from Raft Island, ran down Carr Inlet, and rounded the tip of Fox Island **(Illustration 19a)**. We beat our way up through the Narrows, rounded Point Defiance, and finally guided *Osage* into the yacht club. Patrick and I helped by turning winches under Trumbly's direction, but this was a long way from learning to sail. Nonetheless, it got our feet wet, so to speak. Ever the teacher, Trumbly delivered a rambling impromptu lecture along the way, using *Osage*'s motions to illustrate what he'd been talking about in class for the past several weeks. Most of this blew right past me then; what I got from the trip were impressions.

Osage **(Illustration 19)** was a masterpiece. The browns of her teak and mahogany topsides stood out against the blue-gray of the sea like topaz in a velvet box. *Osage* cut through the chop like a porpoise and shed spray from her oiled decks like drops off sealskin. I could feel gusts of wind begin to heel her over, with tremendous force on the sails, but then the heavy ballast of the leaden keel drew her stiffly back, a giant hand enforcing the inexorable laws of physics.

Ultimately, it didn't matter much that we beginners knew little about sailboats. Trumbly walked us through the design process line by line, explaining as he went the relative advantages and disadvantages of different shapes of hull. He discussed sail plans. He showed us how to calculate the volume of the hull and thence the displacement; how to determine the center of gravity, center of buoyancy, and center of lateral resistance of the hull, and the center of effort of the sails; how to calculate the righting arm and righting moment. He gave us rules for how these things relate to the stability and thus the design of a vessel.

When a student finished the lines, calculations, and sail plan for the 40-footer, there was no further requirement in design; he could do as little or as much of it as he wanted. With no exceptions that I am aware of, however, students chose to continue with design. Not one of us had begun the program without at least the germ of the idea of building his own boat, be it a fishing dinghy or a cruising sailboat. Furthermore, inherent in design is a feeling of the godlike power of pure creation. It is magic to start with a blank sheet of paper and, with a few sweeps of the pencil, lay down an idea that might one day carry you over the water.

Just as with the hand tools used in building boats, we beginners began to accumulate the tools used in designing them. Trumbly lent us a mold to cast leaden drafting ducks, weights that hold in place a grooved plastic rule called a spline, used to draw curves. We scavenged scrap lead wherever we could find it (battery terminals, balancing weights from car wheels, the occasional windfall of a lead pipe), melted it in a can on the kitchen stove, and poured it into the mold. We attached a bent section of stainless-steel welding rod for the neck of each duck and painted each with a thick coat of glossy white enamel. Finally, we covered the bottom of each duck with green felt. Other tools we purchased in the school store—spline, architect's rule, triangles, and a set of drafting instruments.

What I decided to design for myself was a small sailboat, something I could realistically hope to complete while I was at Bates. Unfortunately, I had not even a vague idea what the thing should look like. Therefore, I chose to follow the cardinal rule of the beginning designer: when in doubt, imitate. In a sailing magazine I found a picture of a boat I liked, a popular British class sailboat called the GP 14 (general purpose, length 14 feet). Class sailboats are raced only against other boats of the same design. The intent of a racing class is that, since all the boats are nearly identical, the outcome of a race will be determined more by the skill of the sailors than by the superiority of individual vessels. A popular class boat is likely to be an adequate design, or it wouldn't have become popular.

I can honestly say that I did not copy the lines of the GP 14; indeed, I didn't even have the lines. What I did copy were the parameters of the boat: the length, beam, height, cockpit size, and general shape. I made slight modifications here and there, such as in the curve of the stem, height of the sheer, and angle of the sides. I also slightly reduced the mast height and sail area. I reasoned that if I didn't deviate too much from the parameters of a proven vessel, my design could not be a total disaster. I made changes to the GP 14 design for their own sake, with little idea what their effects would be, because I wanted to be able to claim what I drew as my own, and unique. And my design was unique, though I must admit that when I finally built the boat, it bore an undeniable resemblance to a GP 14.

⌇

My sailboat was to be a hard-chine, V-bottomed boat constructed of plywood. The term "hard chine" simply means there is a sharp angle between the sides and the bottom, rather than the sides curving into the bottom, in which case the boat would be "round-bilged" or "round-bottomed." A hard-chine design is ideal for plywood construction, because each side of the boat can be attached as a single long sheet, as can the two halves of the V-bottom.

A problem arises in designing a V-bottomed plywood hull that is encountered in no other aspect of design. The two halves of the bottom run nearly flat aft, but curve at first gently and then severely toward the stem. The problem is that, although plywood can be bent into a curved shape, it can't be bent in two different directions at once. As it curves toward the stem, a sheet of plywood will not lie flat on the bottom "futtocks," or members, of the forward frames, but instead will bow out from them. These futtocks must thus be slightly curved along their outer edges to allow the plywood to lie flush.

After drawing the lines for a V-bottomed hull, it is possible to determine the proper curvature of the forward bottom futtocks. In the jargon of boat design, this is called "developing the lines." The principle involved is that the bottom plywood follows the surface of an imaginary cone as it bends around toward the stem.

According to Trumbly's method for developing the lines, once you complete the lines drawing, you mark two separate points to the right of the lines. These points represent the apex of an imaginary cone, one relative to the profile plan and the other to the half-breadth plan. You draw several lines radiating from each of these points back to selected reference points in these two plans, and eventually this allows you to determine the correct curvature of the forward bottom futtocks in the body plan, thus arriving at the true shapes of the frames. At best during this process, you have to fuss with the lines repeatedly until the three plans in the lines drawing are mutually consistent. At worst, you discover that the placements of the initial points were wrong. You must then select new points and start again.

Trumbly was uncharacteristically vague about where to place the initial apex points. He understood instinctively, through long experience, where to place them; if you asked him, he'd show you where to place them. But he either hadn't worked out a general principle or hadn't learned to explain the technique adequately, and thus couldn't really teach anyone else how to do it. This must have driven him crazy.

It was in conjunction with teaching us how to develop the lines that one of my classmates, Dan Hubley **(Illustration 35)**, developed an intense grudge against Trumbly. Dan was one of three ex-military men who overlapped the course with me. Dan grew up in a coal-mining town in West Virginia and had opted for the relative safety of the military over the mines. He must have enlisted at a young age, because he was in his early forties when I knew him, and after 20 or more years in the Navy he'd retired at the rank of chief petty officer, the highest non-commissioned rating.

Dan's wife was a telephone operator, a career she'd had most of the time Dan was in the Navy, work she could do as she followed her husband around the globe. The Hubleys lived in a small house in Gig Harbor with their teenaged daughter. Dan was using the GI Bill to attend Bates, with the goal of eventually opening a small boatbuilding shop.

Dan didn't mingle much with the rest of us. He didn't often go up to the cafeteria for coffee breaks, but instead stood outside near the T-38 drinking coffee from a thermos while he smoked. His dark hair cut short in military style was beginning to show a bit of gray. An aging warrior out to pasture, he was also putting on some weight—but out of long habit, he told me, he still did forty pushups and a hundred sit-ups every morning.

Sometimes I carried my coffee down from the cafeteria and talked with him. The first fall I was at Bates, I took a course in celestial navigation taught at Tacoma Community College by a retired naval officer. It turned out that Dan had, among many other jobs in the Navy, worked as navigator. Even though Navy ships had loran and radar, regulations required that a navigator also take solar sightings and plot sequential positions manually as a ship neared land.

"The Navy believes in redundancy and backup," Dan told me. "They are very good at not underestimating the capacity of sailors to fuck up."

Dan had a haunted look, with dark shadows under his eyes. When I knew him, it was only a few years since the US had dragged its ass out of Vietnam. Dan didn't talk much about the war, but once, in an unguarded moment, he said he thought it was the worst thing that had ever happened to the US; it was a disgrace.

"Most people in the military didn't want that war," he said. "Who knows better than a professional soldier or sailor the price you have to pay?" I asked him if he'd gone to Vietnam.

"Yeah, I did a tour running riverboats. It was pretty routine. Day and night, at irregular intervals of thirty seconds to a couple of minutes, we lobbed grenades into the water. It discouraged enemy frogmen. In spite of it, I saw a lot of kids your age die." I began to understand the haunted look. I suspect Dan had kept a lot of kids my age alive, too.

For some reason, Dan took a liking to me and offered fatherly advice. For example, there was the time Shirley came to visit me. Shirley was the secretary in the business office of the boatyard where I started working seven or eight months after I got to Bates. On one occasion, when she expressed an interest in Bates, I invited her to come up for a tour. She showed up on the day we'd arranged, over-perfumed and overdressed. I guided her around the shop and walked her up the gangway to the T-38. After she'd thanked me and gone, Dan took me aside.

"This is none of my business," he said, "but I have to tell you, you can do better than her. I know her kind; I've seen a hundred like her. She's a camp follower, got knocked up by some soldier who's long gone, and she's looking for another sucker to support her and her brats. Whores like her are a dime a dozen. You should forget about her."

Although I didn't have any romantic interest in Shirley, I finally realized she might have had some in me. Although Dan's assessment seemed unaccountably bitter, it was also deadly accurate. I later learned that Shirley was, in fact, divorced from a soldier who had been stationed at Fort Lewis, and she was single-handedly raising two small kids.

On another occasion, Dan was encouraging. When I entered Bates, I'd had a small amount of experience as a journalist, writing local-interest articles for the daily newspaper in Kodiak. In Tacoma, I'd started writing boatbuilding articles for *National Fisherman*; my first was a long piece on the Martinolich boatbuilding dynasty. I didn't think of myself as a journalist; my only goal in writing was to bring in some cash, just as one might drive a beer truck on weekends. Dan, however, thought differently.

"You've got a real talent in being able to write," he said. "Most people couldn't put together a sentence if you held a gun to their head. If I could write like you, I wouldn't be wasting my time in this fucking school, that's for sure. Anyway, I liked your article, and I hope you continue to write."

So I got along fine with Dan, which made it all the more difficult to understand the raw loathing he came to have for Trumbly after the incident of developing the lines.

The navigator in Dan didn't like the imprecision in Trumbly's method, so over the course of a few evenings at home, he came up with a technique that absolutely eliminated the imprecision. Dan stuck pins into the initially estimated apex points and into the relevant intersection points in the lines drawing. He looped thread back and forth from the apex pins to the pins in the lines. He could then move the apex pins around, simultaneously moving the webs of thread. This allowed him not only to determine the optimal positions of the apex points, but also instantly to adjust points in the profile and half-breadth plans. Basically, he converted an inexact and lengthy process into something that could be resolved mechanically in short order.

Dan was justifiably proud of this clever invention, but when he demonstrated it in class, all Trumbly said was, "Look, I've seen people do this before. You can use it if you want, but I don't want you showing it around. I want everyone else to use the method I taught them."

I was astounded. This was the only time I saw Trumbly act other than excited about an innovation in boatbuilding, and the only time I ever saw him behave in what seemed to be a mean-spirited manner. No one understood his reaction. Maybe Trumbly was jealous he himself hadn't discovered Dan's ingenious solution, but this didn't make sense, as he generally gave credit where credit was due. Perhaps he knew some flaws in the method, but this didn't make sense either. Usually when someone suggested a flawed method, he outlined exactly what was wrong with it. Maybe there was just bad chemistry between him and Dan that had nothing to do with boatbuilding.

Whatever the case, Trumbly's response just wasn't Trumbly-like. Most of us chalked up the incident to his having a bad day, but Trumbly's words planted in Dan a resentment that grew with time, like a noxious weed.

"That arrogant prick," he'd say, "he's just like a lot of officers I knew. They'd show off and take chances in order to impress their men, but officers like him were the kind that got people killed."

This wasn't a fair comparison, of course, because there's a world of difference between boatbuilding and combat. Some of Dan's comments were even more personal. Once after Patrick and I had visited Trumbly on Raft Island, where he and his wife Etta lived in a big A-frame house situated in tall hemlock woods overlooking the shore **(Illustration 41b)**. I happened to remark to Dan what a beautiful spot the Trumblys had.

"The only reason Trumbly's got a pot to piss in is because the sonofabitch married that poor woman for her blueberry farm," said Dan.

There was an element of truth to this statement, but it was all twisted by Dan's bitterness. Etta had indeed owned a blueberry farm, but she and Trumbly had first met working in a shipyard during the war, Etta as a welder and Trumbly as a welding lead man. Twenty years later, both divorced, they met again by chance and became devoted to one another. Trumbly had a good, steady job; he didn't need a blueberry farm.

Dan once told me his favorite bar in the world was at the port of Callao in Brazil. The bar overlooked a slaughterhouse, and he could sit with his drink and watch as workers killed steers with a sledgehammer and then hooked them upside down by the heels to a cable that ran them up an incline to the butchering floor. After Vietnam, maybe that was a soothing sight.

Yes, Dan'd seen some ugliness in his life, and I think what he was doing there at Bates was trying to recover from it. Dan was a good man, and the thing he had for Trumbly was just bits of the ugliness working their way to the surface and emerging like shrapnel from an old wound.

⌒

Students came to Bates Boatbuilding from various places and backgrounds. Armen Melkonian came from a truck-driving family in Eureka, California. Lenny Viola and his wife had spent two years during the 1970s in Kabul, Afghanistan, attracted there to an unlimited and virtually unrestricted supply of drugs. A guy named Dave had spent a couple of years in Paris's Left Bank, probably for similar reasons.

The unofficially acknowledged star of the class, however, was a local guy named George Chambers (**Illustration 20**). I say unofficially, because there were no grades and thus no rankings, competitions, or awards. George stood out because he did good, fast work, and Trumbly depended on him. Whenever an especially difficult job came up that baffled the rest of us, Trumbly'd say, "I'll get Chambers on it." George's brother Tim was in the program at the same time, but everyone knew Trumbly meant George when he said Chambers. No one begrudged George this unspoken status; he'd earned it.

With dark, curly, shoulder-length hair and a classically proportioned face, George looked like the young Lord Byron; he had the sort of handsomeness that drove women crazy. Unexpectedly, he was quiet, good natured, and unpretentious. He went about his work and left other people to theirs. He was so even-tempered, in fact, that I thought there must be some serious, hidden personality flaw that would surface once I got to know him better. If there was, I never saw it.

George and Tim were among several sons of a prominent Tacoma surgeon. The eldest one I met had short hair and was involved in some kind of business; he seemed to be the acorn that had fallen closest to the oak. The other three I met were wild, engaging freely in alcohol and any other mind-altering substances that came their way. The second eldest had blond hair halfway down his back and looked like a rock musician. George and Tim were the youngest of the sons; they were not long out of high school and hadn't yet flown the coop.

The Chambers family's main residence was a large, elegant, ranch-style house surrounded by wooded, landscaped acreage in a good neighborhood in South Tacoma. Once when Tim and I happened to be in the neighborhood—in fact, I think he'd given me a ride after school to pick up yet another replacement porthole glass—he invited me to dinner. As we entered the house, I was surprised to find George playing something classical on a grand piano. I remarked that I hadn't known he played the piano. "Well," he said, "I'm just learning."

Dr. Chambers might have tried to keep his wild sons under tight rein so as to mold them as closely as possible in his own image, but he'd taken another tack. To preserve his equanimity, he'd had a bunkhouse constructed adjacent to the main family dwelling. This was where the sons lived. To maintain unity, however, Dr. Chambers had decreed that the family would take evening meals together in the bunkhouse.

Dr. Chambers was a serious amateur chef. It was more than a hobby with him; it was an obsession. He was notorious for pestering the top professional chefs in the best restaurants of Tacoma and Seattle until they divulged their secrets: Chinese, French, Scandinavian, Mexican, none was immune. When I arrived at the table, the Doctor was just setting out a massive salmon cooked in what he announced was French style. I don't remember much else about the dinner, except that it was the best I'd eaten in a long time—though certainly no chef ever had a more willing and uncritical group of test subjects for his culinary experiments.

Chapter 8
Descent to the Flats

It has been stated time and again that, in general, there are two classes of these boat-building hearties: those who contract and build as a means of obtaining a livelihood, and those who build solely for the gratifying diversion and round of pleasure that follow this fascinating pastime.

—Cliff Bradley, *Building the Small Boat*

I finished the design for my 14-foot sailboat before the end of my first fall at Bates. It was a measure of how much I'd learned in less than four months that I'd not only drawn the lines of a boat but also knew how to build it. All I needed to begin construction was time and a place to do it.

Trumbly granted me a few days of school time to loft the boat at Bates, which I did soon after finishing the design. I also began making patterns, using plywood scavenged from the shop and time scavenged from coffee breaks and lunch hours, but there was no way Trumbly was going to let me build the boat at Bates. It was his longstanding policy that students did not focus on personal projects during their first year. Furthermore, given the urgency of finishing the T-38, it was highly unlikely that he'd allow me the time during my second year either. It was clear that if I wanted to build the boat, I'd have to do it on my own time and somewhere else.

The somewhere else materialized in November. Earlier in the year, back in Kodiak, I'd acquired a girlfriend named Mary Ann. When it came time for me to depart for boatbuilding school the following August, we made no commitments to one another. Ours was a tense relationship from the start, and we avoided any discussion of whether we would stay together. We simply parted when I left for Tacoma.

During the fall, Mary Ann had peregrinated to Washington state and was working for a candle maker in Port Angeles. We reestablished contact and visited one another a couple of times in both directions. Finally we decided to get back together, and in November she moved to Tacoma and enrolled in a piano tutorial at Tacoma Community College. She didn't like my apartment on Tacoma Avenue, so we rented a small, two-story cottage in an alley near the top of McCarver Street in Old Town. Even in those days, the cottage was a good deal at $75 a month. Better yet, there was space alongside to build a sailboat.

In January, I purchased a 14-inch band saw and moved it into the kitchen. I thought this only fair, since Mary Ann had bought an upright piano and moved it into the living room. To avoid the raw winter weather, I began shaping the boat's stem and keel in the kitchen as well. Trumbly occasionally remarked wryly, and presumably on the basis of his own experience, "Women and boats don't mix," but what I embarked upon that January was an egregiously flawed test of this proposition. (Yeah, I know, not correct politically or in any other sense, but Trumbly was from a different era.)

Mary Ann took a job as waitress at Bimbo's Italian Restaurant on Pacific Avenue. Bimbo's sat next to a skid-row mission where homeless people could get a hot meal if they endured a long sermon first. I know this because I attended one of these sermons, out of curiosity. Whatever their problems, the street dwellers did not lack spunk. Every time the lay preacher mentioned moral degradation or Jesus, there'd be derisive whistles and mutters of "Blow Jesus out your ass," and "Just give us our food."

Despite its location, Bimbo's was the best-kept culinary secret in Tacoma. The flavor of the spaghetti sauce exploded into your nasal passages like pure gastronomic cocaine, and once you tasted it, you had to have it at least once a week. The secret was garlic. Bimbo's used mounds of it in each batch of sauce, and peeling garlic was one of Mary Ann's jobs during slack periods.

Friday evenings, I'd go down to Bimbo's for my spaghetti fix. I had to eat fast, though, because Mary Ann didn't like me there while she was working. She was especially irritable if there were Gypsies in the place. Typically, a family of seven or eight would occupy a table, order a couple of sodas, sit there for an hour, demolish the place settings, and leave without tipping—although the tip on a couple of soft drinks wouldn't have been much anyway. I think the Gypsies went there to get warm.

Every time I hear Für Elise, it draws me back to that melancholy winter in Tacoma, with rain drumming on the roof of the cottage and Mary Ann practicing Beethoven's haunting melody on her second-hand piano. Things should have been wonderful, but they weren't. I wish I could blame this on the Gypsies, but I can't. The problem was that Mary Ann and I simply could not agree on anything. When she wanted to zig, I wanted to zag. If I wanted to ascend, she wanted to descend. If we wanted to take a bus across town, we argued bitterly about which bus it should be and then ended up on separate buses. None of this boded well for a long or happy relationship.

Tension escalated to a breaking point that came early in March. Mary Ann got home from the restaurant late one Saturday night. I remember it was a Saturday because she wanted to watch Saturday Night Live, and I didn't. I hadn't grown up with a television and never watched it; to me, it was just noise. She turned on the television anyway, which was her prerogative because it was her television. About 10 minutes later, I snapped. I picked up the television, wrenched its cord from the wall, staggered out the front door with it, and hurled it into the alley. It imploded with a pop and a burst of flashes.

Strangely, I watched myself wrestle the television as though I were outside my body, an actor playing to an audience of two. It was highly cathartic melodrama, but it was also the wrong thing to do. Within a week, Mary Ann had sold her piano and was on a bus back to upstate New York. She said she was going to visit her mother and promised she'd be back in a few weeks. We both knew she wouldn't be back.

Before hurling someone else's television, it's always a good idea to consider the consequences. The major consequence for me, other than the emotional fallout, was financial. During the autumn, I'd lived primarily on proceeds from the sale of *Devil's Paw* and my pickup truck in Kodiak. Recently I'd splurged on a band saw and boat materials, and I was nearly broke. Instead of having to pay half the rent, I was suddenly faced with all of it. Unfortunately, I didn't have all of it.

$\backsim$

A job materialized soon after the hurling of the television, as though it had been written into the script of the melodrama. One afternoon at Bates, when I was out working on the T-38, a new, green MG convertible pulled into the courtyard and parked. A blond, well-groomed, bearded guy in his mid- to late twenties got out and entered the shop, where he conferred with

Trumbly. Later that same afternoon, Trumbly drew Patrick and me aside and told us that Knapp Boatbuilding, a small, family-owned boatyard, wanted to hire some students and that he'd recommend us for the job if we were interested.

The next day after school, we loaded our toolboxes and carpenter's coveralls into Patrick's pickup and drove to the Flats. There was a considerable difference between viewing the Flats from the verdant residential hillsides of North Tacoma and working there. From afar, the Flats had been an abstract concept, a battlefield of industry almost romantic in scope. Close up, the wastage of industry was evident wherever one looked. The air had a hazy, cream-colored tinge like a poisonous cloud, which in fact it was. If you left a car overnight, droplets of clear, yellowish organic-chemical gunk condensed on it; the sun did not evaporate the droplets but merely turned them gummy. Unhealthy trees and patches of scrofulous weeds struggled for existence in the chemical milieu of vacant lots, and only bacterial sludge and stunted algae seemed to survive in the noxious waterways.

I simply had not made the connection that this was the type of environment Bates Voc was training me for. Stepping out of Patrick's truck with my toolbox, my first thought was, "My god, what have I gotten myself into?" This was followed by panic and fleeting images of the Devil, cancer, war, and concentration camps. It was too late, though, to back out without admitting I'd made a huge mistake coming to Tacoma, to say nothing of having to face Trumbly with an excuse for not taking the job, such as "It's too nasty down there on the Flats."

Knapp Boatbuilding was sandwiched between Marine View Drive, which skirted the base of the bluffs along the far side of the Flats from Tacoma, and Hylebos Waterway, which ran parallel to the Drive. Knapp occupied a big, rectangular, two-story wooden building with a low-peaked roof. Facing the road on ground level at the front of the building was an attached business office, with parking for a few cars in front. An alley ran along the east side (the left side as seen from the road) of the shop to a graveled lot in back. At the far side of the lot was the smaller shop of S & S Boatbuilding, a yard that built aluminum fishing boats, and beyond that, the waterway. In the wall of Knapp's shop facing the back lot were two massive doors that allowed egress for finished boats.

Knapp Boatbuilding was surrounded by boats and boat paraphernalia. Across the alley was Hylebos Boat Haven, a marine haul-out, storage, and repair facility where a couple acres of boats of all kinds—aluminum seine skiffs, sleek fiberglass sailboats, old wooden power yachts—lay blocked up. The Boat Haven had a mobile cradle for shuffling boats around the storage yard, and to and from the waterway. A cluster of boathouses lined the waterway behind Knapp and the Boat Haven. On the side of Knapp opposite the Boat Haven was an undeveloped lot where knee-high weeds partly hid an assortment of maritime debris—a decaying fiberglass deck mold, empty drums, a rusting steel mast, broken engines.

Patrick and I entered the front office and introduced ourselves to Don Knapp, the company's owner and operator **(Illustration 21)**. Don was a burly, florid man in his late fifties. With a tonsure of graying brown hair, he looked like a cross between a clean-shaven Santa Claus and a former professional linebacker. He'd outdone Trumbly in terms of fingers; several of his were missing.

Don eyed us skeptically and then led us on a tour of the boatyard. The shop proper was a cavernous space 70 feet long, 45 feet wide, and two stories high. Along one end was a small second-floor loft accessed from the shop floor by a wooden stairway. An interior wall along the east side of the shop separated it from two floors of accessory rooms. The ground floor contained a series of small workshops. The upper floor was partitioned into a lunchroom, a bathroom, a long storage space, and a construction office that had a sliding window overlooking the shop floor.

Don ushered us into the construction office and welcomed us to Knapp Boatbuilding. "I'll tell you guys right off that I wasn't in favor of hiring you," he said. "You're inexperienced, and I don't trust your work. But my son Chuck seems to think it's a good idea. This job's behind schedule, so I'll go along with it for now, but don't expect journeyman's wages, because you aren't journeymen."

Don had been gazing distractedly into the shop, and he suddenly slid the window open with a bang and thrust his head out. "For Christ's sake, you're dragging that cord through the deck coat," he shouted to someone. "Yes, you! Wake up and watch what the hell you're doing!" He slammed the window shut.

"The main thing to remember," he continued, as though there had been no interruption, "is that you're here to work, not screw around. When I pay you an hour's wage, I expect a solid hour's work, the best and hardest you can give. If you do that, we might get along. If not, I'll fire you."

Don took us down to the secretary to fill out employment papers and then relinquished us to Chuck, the foreman, who turned out to be the bearded guy I'd seen the previous day at Bates. After punching our time cards—a novelty for me, as I'd never seen a punch clock before—we followed Chuck into the shop to start work.

The boatyard was not a quiet place; the air was saturated with the sounds of grinding, pounding, sawing, tapping, clanging, drilling, bumping, and shouting. Talking was difficult, which was okay, because we soon learned that Don frowned on talking. He figured that if you were talking, you weren't working.

The shop was a microcosm of the Flats. Wood scraps, sawdust, twisted fragments of resin-coated fiberglass, spatters of hardened resin, and paint stains littered the floor around the boat. A thin layer of the white dust of ground fiberglass coated everything and everyone. Light from outside highlighted the particle-laden atmosphere in the shop.

The vessel under construction was a 53-foot custom charter-fishing boat named *Macs' Effort*. Incredibly, the name was grammatically correct; the boat had been commissioned by two brothers, Bob and George McPherson, of Westport, Washington. At the boatyard, the McPhersons were referred to as the "owners," as opposed to Knapp, who was the "contractor." Strictly speaking, the McPhersons didn't own the boat—a bank owned most of it and would for a long time.

⌒

The fiberglass hull of *Macs' Effort* (**Illustration 22**) was a stock 48-foot, V-bottomed planing model designed by Lynn Seynour of Seattle and built by the Uniflite Company of Bellingham, Washington. Uniflite had a lot of experience building durable fiberglass hulls; during the Vietnam War it produced a thousand or more 31-foot river patrol boats and 36-foot landing vessels for the Navy. Knapp had lengthened the stock hull to 53 feet by adding a 5-foot section just forward of the transom. This was the first Uniflite hull ever lengthened in this fashion, and everyone involved—the owners, the Knapps, and Uniflite—were nervously curious how it would perform. *Macs' Effort* was to be powered by two 425 horsepower V-8 turbocharged Detroit Diesels that would propel the vessel, its crew, and as many as 25 day-passengers at upwards of 20 knots. This speed was necessary in the highly competitive charter sportfishing business, where the first boat to reach a school of fish tended to catch those fish and thus keep the clients happy.

In profile, the vessel was apportioned roughly in thirds, with the raised wheelhouse occupying nearly the forward third, the passenger cabin the middle third, and the low, open fishing deck the aft third. Fifteen feet in beam and nearly 16 feet high from the bottom of the forefoot to the top of the wheelhouse, *Macs' Effort* was approximately the size of an 800 square foot mobile home with a partial two-story addition on top. But a lot more was crammed into the *Effort* than one finds in a mobile home. In addition to the massive diesels and fuel tanks, and all appurtenances necessary to direct, propel, and fish the boat, were a galley, three staterooms, and two heads, complete with showers. With dinettes and settees that converted to bunks, the *Effort* could sleep 16.

In our inexperience, it seemed ironic to Patrick and me that Knapp Boatbuilding had not built the hull of the vessel. However, this was a common practice for smaller yards, dictated by economy of scale. Knapp could buy a bare fiberglass hull from a company that produced them by the hundreds much cheaper than they could build one.

A layman might go so far as to contend that Knapp Boatbuilding did not really build *Macs' Effort* at all, arguing that the hull is the defining aspect of a boat. In this view, constructing the hull is equivalent to constructing the boat. However, this is not the case. The hull of a vessel of this type comprises a low fraction of the construction costs in terms of time and materials. The hull is simply the shell of the vessel. Hull framing, bulkheads, and decks must be constructed inboard. The wheelhouse, galley, heads, staterooms, storage compartments, fish hold, lazarette, chain locker, engine room, and compartments for fuel and water tanks must be framed in, sided, and in some cases made watertight.

All of this is still just the shell, really. Cabinets, counters, benches, bunks, ladders, railings, and appliance compartments must be built, and built well. Anything that can be grasped with the human hand must be strong enough that a seasick passenger will not pull it loose as he distractedly fumbles his way toward an appropriate place to puke.

When the framework is in place, the "ics" must be installed: the mechanics, electrics, and electronics. The "mechanics" include the engine, engine cooling system, exhaust system, propeller shaft, rudder, deck winches, anchor winch, steering assembly, bilge pumps, and engine-powered hydraulic systems with the mountings necessary to hold all these things in place. Installation of the mechanics requires very exacting work. The propulsion system, for example, must line up perfectly without any sources of vibration that might cause it to self-destruct.

The "electrics" are the wiring of the vessel and the electrical devices connected to the wiring. The wiring of a boat is a hybrid between automobile wiring and house wiring. Current flows from the engine generator to the battery, and thence both back to the engine and to many of the same fixtures and appliances found in a house. *Macs' Effort* had three electrical systems with 96 circuits among them: 12 volt DC, 24 volt DC, and 110 volt AC. The vessel carried an auxiliary diesel generator for when the main engines were shut down or could be plugged into shore power when tied up at the dock. An extensive alarm system provided audio and visual warnings in the wheelhouse of engine malfunctions, transmission overheating, low exhaust-water output, high bilge water, and failure of the bait-tank circulating pump.

The wiring of a boat requires special attention because of the great potential for it to get wet and short out. There are other peculiarities as well, for example the phenomenon of stray-current corrosion, whereby improperly insulated or grounded wiring can cause metallic parts of the boat to dissolve and fall apart in the saltwater environment.

The "electronics" are the navigation and communication devices. In *Macs' Effort*, these included radar, loran C, a graph depth recorder, and a flasher; VHF, sideband, and CB radios; an automated radio direction finder; the compass and autopilot; and a hailer/intercom system.

The systems in a boat are interconnected and interact with one another. For example, the navigational instruments and compass are linked to the autopilot, which is linked to the steering system, which is linked to the hydraulic system, which is linked to the engine, which is linked to the electrical system, which connects back to the navigational system and compass.

A boat also needs plumbing, and lots of it. Plumbing connects a freshwater tank to sinks, showers, and toilets; sinks and showers to external outlets; and toilets to a sewage holding tank. Other plumbing allows pumping of the bilge, seawater circulation through the bait tank, and circulation from the engine to whatever cooling system is used. The pipe connections in a boat must be especially sturdy so as not to vibrate loose under seagoing conditions and cause the vessel to be flooded with sewage or seawater.

After all this, there is finishing. Finishing involves making everything pretty and functional, the final touches. Through tradition and necessity, finishing is of higher quality in a boat than in the average house. Fine hardwoods are used for visible wooden surfaces, such as doors, molding, and trim. Bulkheads are covered with vinyl or hardwood-veneer paneling. *Macs' Effort*, for example, had teak veneer on the bulkheads and louvered teak doors on the staterooms. Because boats are full of curves and angles, finishing pieces must be exactingly cut and fit; thus the labor of finishing a boat is intrinsically greater than that for a house. In the last stages of finishing, ceiling spaces are covered with paneling or fabric held down by removable hardwood molding, to allow easy access to the wiring. Inside decks are covered with paint, linoleum, teak parquet, or water-resistant carpet, depending on the owner's budget. Furnishings are installed: toilets, sinks, towel racks, stove, microwave oven, refrigerator, and the like.

It required a lot of work, then, to convert the bare hull of *Macs 'Effort* into a boat, which is why it is accurate to say that Knapp Boatbuilding built the boat. A modern boat is a hybrid between a house and an automobile, but is more complex than either. Designed to function stably in a fluid environment, it has more in common with an airliner than anything else.

A modern boatbuilder knows how to estimate and draw up a bid, obtain financing, construct a hull from a set of lines or modify a stock hull, build the guts and superstructure, install the complex systems, and make everything pretty. He knows where to obtain quality materials at minimum prices. He is familiar with Coast Guard regulations governing the construction of a vessel for hire and understands the complicated certification and documentation process. He has contacts in the industry so that if an unusual problem arises, he can seek advice from someone who's encountered it before. Finally, he knows competent carpenters, fiberglassers, electricians, plumbers, and mechanics he can count on to come to work when he signs the contract to build a boat.

When Patrick and I arrived at *Macs'Effort*, the cabin and wheelhouse were externally complete, and most compartments inside the vessel had been framed in and sided. Our first job was to build bunks and benches in the staterooms. Chuck described what he wanted, and then left us alone. Since we were to work separately in different staterooms, we competed to see who could finish a bunk faster, with the stipulation that we could not cheat by sacrificing quality. Over the following weeks, we worked every day after school and all day Saturdays. We soon came to know our co-workers and what they did at the yard.

As contractor and construction supervisor, Don oversaw the ordering and delivery of materials and the hiring and firing of workers. He tracked the construction as a whole, verifying that it adhered to the plans and met the terms of the contract. In doing so, he interacted with the owners, one of whom, Bob, made periodic tours of the vessel to verify Knapp was not cheating him. Additionally, Don negotiated any changes Bob wanted that departed from the contract.

As foreman, Chuck interacted directly with the workers, assigning them where they were needed. He roamed over the boat, checking work and offering advice when problems arose. When he wasn't doing this, he occupied himself by helping install the mechanical systems. In principle, Chuck was the traditional link between management and labor. A good foreman wins the trust of his charges, so that they will function as a team under his direction and confide in him when a serious mistake occurs. However, because Don and Chuck were father and son, Don often ignored the chain of command, micro-managing the work with frequent shouts from his office window.

Chuck was normally respectful of Don, at least in public. One afternoon, however, finally fed up with Don's tirades, he flipped his father the bird from down on the shop floor. To the rest of us, who were terrified of Don, this was like flipping off God, and noise in the shop slackened as we awaited the outcome. Don's response was much like you might expect from God. We heard an angry bellow, a chair scraping back and hitting a wall, a door slamming, and heavy footsteps pounding the length of the second floor, sounding like an enraged water buffalo. Don appeared at the head of the stairway running from the shop floor to the loft.

"CHUCK! That's the first and last time you'll do that to me, by God! I'm not too old to kick your ass from hell to high water. I want you in the office, NOW!" After a few minutes to cool off, Chuck went up to the office, and we heard a violent, muffled argument.

Gary was a journeyman boat carpenter in his mid-thirties, swarthy, short, and stocky, with a handlebar moustache. He'd worked on *Macs' Effort* from the start and had done much of the structural carpentry. Now he was undertaking finishing carpentry, anything that would remain visible, such as railings, molding around tables and counters, hardwood caps on the bulwarks, and cabinet facings.

Griz, the primary fiberglasser on the job, was now occupied with finishing-stage fiberglassing and gel coating. In his early twenties, he was over six feet tall, lanky, and strong as a bear, hence his nickname. He wore his blond hair in an Afro. Griz was a dedicated quipster. For example, if you called him a wiseass, he might reply something like, "Yes, that's true; I leave no ass unwise." There was very little he couldn't fashion a quip for.

The electrician was Pershing Haynes, nicknamed Perk. He was a tall, burly man in his sixties. With a pink face and a full head of white hair, he looked like an overgrown leprechaun. An old crony of Don's, he was a friend and advisor to the Knapps. During Don's tirades, Perk would continue his wiring, shaking his head almost imperceptibly, with a slight smile on his face. Already retired from a previous job, he took wiring jobs only when he felt like working.

Chris was a thin, pimply-faced kid just out of high school. His job was to lend a hand where it was needed. He fetched things, held things, cleaned up the shop, and did rough fiberglassing in places it wouldn't be seen. He intended to work until he could get into the Navy, which was his life's dream. No one thought Chris was too bright, but maybe he just seemed slow because he was intimidated.

The mechanic for the boat was Bud Wolreth, who'd installed engine, propulsion, hydraulic, cooling, and steering systems on most of the boats the Knapps had built. He was present intermittently, and I never got to know him. There may have been another carpenter or fiberglasser or two, but those I've mentioned are the only ones I remember.

Then there was Earl. Earl was a plump, soft, round-faced guy in his late thirties, well groomed but nondescript, with thinning brown hair. His role at the yard was unclear. He dressed like a typical yacht yuppie: creased slacks, a boat-logo T-shirt if the weather was hot, a slipover sweater if it was cool, canvas deck shoes. He occasionally helped with small things— he'd pass up a board or straighten out a cord—but he did very little, and nothing dirty. In a place where people climbed around like banshees, covered to various degrees with particulate matter, resin, glue, and sweat, Earl stuck out like a tuxedo in a slaughterhouse.

Various rumors circulated about Earl. He was going to become a partner in the boatyard. He had a degree in aeronautical engineering and was now learning the boatbuilding trade. He was the scion of a mysterious, powerful real-estate woman in Tacoma. He was a crack salesman, wealthy in his own right. Even now, I don't know the truth of most of these rumors. What was clear at the time, however, was that although Chuck was friendly with Earl, Don had no use for him and ignored him entirely.

An incident at the time provided a glimpse into Earl's personality. One afternoon, we noticed that Chris wasn't at coffee break. Chris never missed coffee break.

"It's just like that little prick to punch out early when we're against a deadline," said Griz.

Fifteen minutes after break, however, someone spotted a pair of legs protruding from the chain locker, a compartment near the bow of the vessel, accessible only through a small, vertical hatch. Chris had been doing the brutal job of fiberglassing the inside of the compartment. For some reason, he hadn't worn a charcoal-filter mask and had passed out from the resin fumes. We dragged him by the ankles from the locker. It took him a while to revive, and then he was groggy and uncoordinated. Chuck, worried about possible brain damage, said, "Man, we've got to get him to a hospital."

"Aw, don't worry about brain damage," said Earl. "Hell, in the Navy that'll qualify him as officer material." We all laughed, but Chuck took Chris to the hospital anyway.

After Patrick and I had built a number of bunks and benches, Chuck assigned us to lay down vinyl paneling inside the staterooms. Since we were covering surfaces having complex outlines, every piece of vinyl we applied first required we make an outline pattern. We did this by scribing, trimming, and nailing together plywood strips—a technique we'd learned at Bates. We transferred the shape of the pattern to a big sheet of vinyl and cut out the piece with a band saw. We then glued the vinyl carefully in place. Vinyl is touchy; it can chip when you cut it, will break if you bend it too sharply, and doesn't pull off readily if you position it incorrectly. However, we had no problems once we learned the quirks of the material.

We were thus occupied one afternoon, working together, when we noticed one of the owners, Bob, watching us through the stateroom door. This was Bob's first tour of *Macs' Effort* since we'd been working there. Though we didn't know who he was, we greeted him politely. Bob didn't say a word, but instead fixed us with the same sort of aggressive stare one encounters in biker bars, like he had a brand new Harley and wanted to run it up our asses. He examined our work without comment and then continued his tour.

When Chuck came by later, we asked him who the jerk was and why he was so unfriendly. "That was Bob," Chuck said, "the guy we're building this boat for. He was unhappy when he found out we had Bates students working on it. He claims he's not getting the quality work he's paying for."

"So, does this mean we're off the job?" I asked.

"No, you're still working here. Even though Bob may bitch about it, it's not his decision to make. We hold the contract, and who we hire is our business. He has the option to refuse the boat when it's time to sign off, and then the lawyers can fight it out. But that won't happen, because your work is fine, and because he really needs the boat."

Chuck bent over to scoop up an errant bit of vinyl adhesive with his finger. "You guys just focus on the job and leave Bob to me. He's not a bad guy, but he's a typical owner. He's taking on a huge debt buying this boat, and it makes him nervous."

So we forgot about Bob. Whenever he came by, we ignored him and went on with our work.

Knapp Boatbuilding must have used Patrick and me as guinea pigs to test the feasibility of hiring Bates students. Apparently we passed muster, for a week after we'd started, Chuck asked Trumbly for more students. Since the most experienced students were already working elsewhere, Trumbly announced to the class-at-large that anyone who needed work could talk to Don Knapp. This proved to be a mistake.

One of the new employees was Klarich. There was no question that he'd do Bates proud whatever Knapp threw at him. Another was Tim Chambers, who was competent. Another, however, was Dean Goodrich, a big roll of the dice.

Dean was a full-blooded Native American who'd been adopted at a young age by a couple of white yuppies in Gig Harbor. When you first met him, you expected him to talk with a guttural Native American accent. Instead, he spoke fine, Ivy League English, very cultured, as though his parents had gone to Yale, which they probably had. Dean was still in high school. Very intelligent, bored with traditional classes, he'd taken advantage of an option whereby he could get high-school course credit through training at Bates.

Though he was the youngest student in the boatbuilding class and had been at Bates only a few months, Dean managed to con his way past Don and get hired at Knapp. For the next several weeks, work progressed smoothly for us Bates students. Our tenure at Knapp might have finished smoothly as well, had not Dean run afoul of Bob, the owner, one bright May day.

On one of his routine tours, Bob noticed a bench Dean had framed in the wheelhouse. It was bad luck, really, because in another day siding would have covered the framing, and Bob would never have seen it. But he did see it, and it looked like a job done by someone strung out on acid. With Dean, this was entirely possible. The ends of pieces were split off by bad saw cuts. Screws emerged where they shouldn't. Streaks of hardened glue ran down from wide gaps in badly cut joints. The whole shebang also looked out of square, but this may have been an optical illusion created by the other defects. Whatever the case, when Bob saw Dean's framing job, he went ballistic. Swearing roundly, he shot out of the wheelhouse like a bottle rocket and stomped up to Don's office.

This was the beginning of the end for Knapp's educational-outreach program. Don fired Dean at the end of the day and notified the rest of us we were finished at the end of the week. I made some inane remark to Klarich about how unfortunate this was. Klarich grinned and said, in his enigmatic, whimsical way, "Are you kidding? I wouldn't work here anymore if they paid me."

On the last day, Chuck called me aside and said, "Don't take it personally that we're letting you go. Bob just precipitated a decision that we would've made sooner or later anyway. You guys have helped a lot, and the boat is far enough along that the regular crew can complete it from here.

"But listen, in a few weeks Earl and I are going to start work on a 54-foot, ocean-racing sailboat. We're going to build a plug and a mold, and then finish a demonstration boat. We're going to call it the Aquila line, and if it's as fast as we think it'll be, we'll be able to produce and sell a lot of them. I'm going to start lofting it very soon, and then we'll need someone to make patterns and frames for the plug. After that, there'll be work building the plug and mold, and finishing a hull. Are you interested?"

I'd only been boatbuilding for about nine months, and it seemed preposterous that I had enough training to do what Chuck needed. However, I realized I'd done it all before. I'd made a full set of frame patterns for Trumbly's 7.5-foot, round-bottomed dinghy and built a set of complete frames for my own 14-foot sailboat. Though much larger, Aquila would require exactly the same techniques.

"Sure, call me when you're ready," I told Chuck.

⌐

Tooling up to produce a boat design in fiberglass means constructing a mold from which to cast hulls **(Illustration 23)**—like the mold that Patrick and I had used to make Roberts's little sailing dinghies. To construct a mold, the builder first needs to make a hull to provide the shape for the mold. One way to go about it is to build an actual hull from wood, take the mold from that, and then finish this first hull as a functional boat. However, this is not the fastest way to obtain a mold. What is usually done is to build a full-sized wooden model of the hull. This is called the plug. The plug is identical in shape to the hulls that will eventually be cast one-by-one from the mold, but it will never touch the water. It is weak and temporary; its only function is to provide a form for the mold, after which it is discarded **(Illustration 28b)**.

For clarification of plugs and molds, consider a more familiar but entirely parallel example. Suppose you have a company that makes cheap statues. One day, a concessionaire from Lourdes orders from your company a thousand plaster Virgin Marys. So what do you do? Do you hire someone to sculpt each Virgin individually from plaster? No, what you do is hire a master carver to carve a single Virgin from wood. This is the model, equivalent to a boat plug.

You then polish the wooden Virgin, grease her well to give a non-sticking surface, and coat her with a thick layer of liquid latex rubber. You leave the base of the statue uncoated. When the rubber has solidified, you peel it away from the statue in one piece, starting at the base and working up. The rubber turns inside out when you do this, just as a tight rubber glove turns inside out when you peel it from your hand, so you invert it by pushing it back through itself. This rubber sack, the inside of which has the exact shape of the Virgin, is the mold.

To cast a Virgin Mary, you pour liquid plaster into the mold and let it solidify. You then peel back the mold, and there is the first plaster statue. You rinse off the mold, invert it again, and pour the second statue into it. To complete the order for Lourdes, you repeat this casting process 998 times.

Fiberglass boats are made in a very similar way. Instead of carving the plug for a sailboat hull, the builder lofts the boat, just as he would any other. He makes half-frame patterns from the loft and uses the patterns as templates for the frames. When all the frames are made, he erects them upside down and evenly spaced on a strongback, just as he would do for any other boat built upside down. He then strip-planks the frames—that is, covers them with flexible strips of wood, say 1 inch thick by 1 1/4 inches wide. It is like carvel planking with very narrow planks, except that the strips are edge-glued to one another.

Unlike the mold for the Virgin Mary, the mold for a boat hull cannot be made of latex rubber, for it must be rigid enough to avoid distortion caused by its own weight and the weight of the hull cast inside it. The larger the hull, the stronger and better supported the mold must be. Therefore, hull molds are themselves constructed of thick fiberglass, usually in two (right and left) halves that can be bolted together along a midline flange. When it comes time to remove a hull from the mold, the two halves can be unbolted from one another and pulled away separately **(Illustration 23)**.

In summary, then, to tool up for producing a line of fiberglass boats, the builder must obtain the lines for a particular design, loft it, build a wooden plug, use the plug as a template to construct a mold itself made of fiberglass, and finally use the mold to produce fiberglass hulls.

⟿

By the time of my layoff at Knapp, I'd fortuitously completed of all the elements necessary to set up my 14-foot sailboat. Even while working part-time, I'd made slow but steady progress. Whenever I could steal a bit of time at Bates, I cut out frame pieces and gussets for the frame joints. When I had cut all the members of a frame, I glued and nailed them together, and then bolted on a crosspiece to allow the finished frame to sit upside down on a strongback. In this manner, I gradually accumulated all the frames for the boat.

Now, for a few weeks, I had spare time. Using scrap lumber scavenged on the Flats, I framed in a boat shed alongside my cottage and covered the top and three sides with tough, transparent visqueen plastic. My accountant landlord initially balked at having this eyesore on his property. Since the cottage was located in the densely packed residential neighborhood along McCarver St., he was also concerned about building code violations and noise complaints. I argued, however, that the shed would improve the prospects of his renting the cottage to a future Bates boatbuilding student at significantly higher rent when I left, and he grudgingly acquiesced.

When the shed was complete, I built and leveled a strongback inside it. One Saturday, the little shed hot as the tropics from sun beating through the plastic, I squared the frames on the strongback and nailed them down by their cross members. I plumbed them and tacked temporary battens, or ribbands, fore and aft to hold them in place. I lag-bolted the stem and transom to the strongback in their proper positions and fit the keel into notches cut for it at the apices of the frames. I fastened the keel to the stem, transom, and each frame. In a single day, the shape of the sailboat magically emerged **(Illustration 26)**.

Extending from stem to transom along each side of a hard-chine boat is a longitudinal member of the skeleton called the chine log, the function of which is to support the joint between the side and the bottom. To install each chine log, I notched the angles of the frames to receive it. I then glued and screwed the log to the stem, bent it around and similarly fastened it to each frame, and finally fastened it to the transom. Bending a chine log is tricky, because sometimes it snaps in two at the point of sharpest curvature.

Then there is nothing to do but remove the wreckage, cut another piece, and try again, this time soaking or steaming it before bending.

I had just installed both chine logs successfully when a phone call from Chuck Knapp yanked me out of my boat shed and back into the maw of industry. It was time to start making patterns for Aquila.

Chapter 9
School of hard knocks

GARIBALDI, Ore.—A large wave flipped over a charter fishing boat carrying 19 people off the northern Oregon coast Saturday, killing at least nine, the Coast Guard said. Two people were missing several hours after the capsizing. Rescuers searched near a long, rocky jetty at the mouth of Tillamook Bay, an area known for high waves and swirling currents.

—Joseph B. Frazier, Associated Press, 14 June 2003

Significant changes had occurred at Knapp Boatbuilding in my absence. For one, the company no longer existed. In front of the boatyard was a fancy new sign that advertised "CLK Yacht Crafters." The CLK, it turned out, stood for Charles L. Knapp, that is, Chuck. Through some legal legerdemain, Knapp Boatbuilding had disappeared and CLK Yacht Crafters had coagulated from the vapors of its ghost. Though I never learned the details, Chuck and Earl were either partners in the new company, or it was Chuck's company and Earl was an investor. Anyway, this confirmed one of the rumors about Earl, that he would soon be involved in the boatyard in a significant way.

When I entered the business office, Shirley was no longer there. In her place was a thin, Midwestern-looking woman from Oregon nicknamed Breezie, who was in her early twenties. She wore her straight, blond hair in a long ponytail and had a gaunt look, like she might be anorexic.

The managerial hierarchy had changed as well. Don spent less time in the supervisor's office overlooking the shop and more time in the front office, apparently training Breezie. Chuck spent less time in the shop and more time in the supervisor's office, where he had set up a drafting table. Earl had emerged from the shadows and was actively involved in the final stages of *Macs' Effort*.

Macs' Effort sat where it'd been when I left, and the same full-time crew worked to finish it. It was the dying gasp of the old company, the final loose end to be tied up. The half of the shop floor unoccupied by *Macs' Effort* had undergone a radical transformation. Stacks of materials and shop equipment had been removed. Four-by-eight-foot sheets of 3/4-inch particleboard painted white had been laid down over the concrete floor to convert it to a lofting floor. Sixty feet long and 18 feet wide, the loft looked as pristine as a snow-covered mountain meadow.

Chuck had been busy while I was gone. In addition to supervising *Macs' Effort*, he'd learned lofting through one-on-one instruction from Trumbly and had nearly finished lofting Aquila. This is a good example of the resource Trumbly represented for the boatbuilding industry. It didn't matter who came to him or what they wanted to learn; if it had to do with boats, he was willing to teach.

Chuck immediately gave me my orders. "Aquila is a cutting-edge ocean-racing design, and we want to build it better than anyone has ever done a production yacht before. We've decided that all tolerances will be within a thirty-second of an inch. I've lofted it to that tolerance. Your job now is to help me finish the loft. I want you to figure frame bevels and expand the transom. After that, I want you to make frame patterns and frames, to the same tolerance as the loft."

I doubted this level of precision was realizable in practice, but I supposed trying to achieve it was a useful goal. After dealing with the bevels and transom, I began transferring half-frame outlines from the body plan to 1/4-inch plywood using the nail-head method I'd learned at Bates, and cutting out and fairing the patterns **(Illustration 7)**. It was early summer now, and the shop was a furnace. Just as on the loft at Bates, I worked in a white carpenter's coverall with only my undershorts beneath. Even then, I left a trail of sweat like the track of a giant slug as I crawled over the lofting floor, tacking down nails and lifting patterns.

When the frame patterns were done, I made patterns for the stem and transom. Then I started on the frames, 34 of them. Using a pattern, I traced each half-frame onto 3/4-inch plywood, sawed it out, meticulously faired it, and joined it to its opposite member with gussets and a cross member.

I vaguely remember that, working 20 to 30 hours a week, it took me somewhere between four and six weeks to complete the patterns and frames. This seems like a long time, but Aquila was a monster. Fifty-four feet in length, 14 feet in beam, and over 7 feet deep from base of keel to sheer, the boat was literally as big as a humpback whale.

During this period, Chuck spent at least part of each day at the drafting table, laying out the cabin and interior, making stability calculations, and drawing the sail plan. He and Earl made frequent trips to Bates to confer with Trumbly. Several times, Chuck made slight modifications to the lines. These changes in turn required changes on the loft, time-consuming modifications to frame patterns, and in several instances the rebuilding of frames. Eventually I realized there was something strange going on. It is normal procedure to start with a finished set of lines and loft them straightaway. Why the changes?

So, one day I asked Chuck, "Who designed Aquila, anyway?"

After an awkward pause, Chuck said stuffily, "I really can't tell you that. It's confidential information I'm not at liberty to divulge."

I should have been satisfied with this answer. After all, I was getting paid by the hour. Chuck could make all the changes he wanted, and I could modify patterns and frames until retirement. It wasn't my business. However, if there's one thing I can't stand, it's a secret. As soon as this word attaches to a piece of information, then no matter how useless it is or irrelevant to my life, knowing that information becomes an obsession.

Hence, I didn't let the question about the designer die. I asked obliquely at first, but directly in the end. I'd say, "The designer must have been pretty sloppy, if you have to make so many changes."

Or, "What'd you do, get the design at a discount price?"

Finally, "Come on, goddamn it, I work here. Tell me where you got the lines."

Chuck ultimately relented. "OK," he said, "I'll tell you who designed the boat, but you can't tell a soul. I mean it; there are serious consequences if you do. You see, Earl went up to Seattle and got intimate with a secretary in the West Coast office of so-and-so." So-and-so was a prominent Boston sailboat designer, one of the most famous in the business.

"Somehow, and I never asked how," Chuck continued, "Earl borrowed the lines of so-and-so's latest design from the secretary and had them copied. Now, I'm making enough changes that Aquila won't be recognized. I've lengthened the vessel four feet, modified the sheer, redesigned the cabin and deck for both aft-cockpit and center-cockpit versions, re-drawn the transom, and changed some of the lines. The irony is, I think I'm making it better. We're going to have a faster boat than so-and-so."

Thus the secret was revealed, and I was relieved to learn it involved nothing more than run-of-the-mill industrial espicnage. It was probably a grand-theft felony as well, but ultimately there was something in it for everyone. Earl and the secretary got mutually laid, Chuck and Earl got a set of lines, I got a job, and so-and-so wouldn't have to knowingly suffer the ignominy of having one of his own designs beat itself in a sailboat race.

⌒

During pattern making, a bizarre incident occurred that further indicated things were a little strange at CLK Yacht Crafters. Earl came up to me one afternoon when I was on my hands and knees on the loft. He looked as though he had something to say, and I rose into kneeling position to listen to him. He bent over, so that his face was close to mine.

"We're trying to do a quality job here, and if you don't agree with that, then pick up your tools and go home," he said in a curt, angry voice. This was so unexpected that I could only stare at him openmouthed.

He continued, "We're a team here, and there's only room for team players. I'm one of the captains of that team, and if I give you directions, I expect you to obey them. If you can't accept that, then I want you to leave."

When I recovered from the shock of this onslaught, I got angry. There was nothing wrong with my work, and I knew it. I had a hammer in my hand and felt a strong urge to tap Earl's forehead with it But I didn't. It's not appropriate to hit a crazy person with a hammer. So I said, "Sure, Earl, I can accept that, as long as you don't order me to screw a chicken. But tell me, have I done something wrong?"

"No," he said, "your work is fine, but I just wanted to let you know how things stand." And he walked away.

Later I indignantly related the incident to Chuck, who laughed. "Don't take it personally," he said. "Earl likes to play mind games with people. I'm glad you didn't hit him with the hammer; that would have caused problems."

Chuck must have said something to Earl about the episode, because Earl began needling him the same way. When there was tension or disagreement between them, Earl would stare at Chuck and say, "Chuck, just try to do a good job." It was meant to be humorous because of the obvious irony—Chuck was supremely meticulous and knew far more about boatbuilding than Earl. With repetition, though, the remark lost its humor and seemed to intensely annoy Chuck.

While I worked on Aquila, everyone else worked hard to finish *Macs' Effort*, which still filled the other half of the shop like a white elephant. And a white elephant is just what it'd become, as well as a thorn in the side and an albatross around the neck. The boat was all of these combined into a grand mixed metaphor made of fiberglass. Everyone was sick of it; Don was losing money on it; Bob was pissed because it was still in the shop with the fishing season well underway; Chuck and Earl saw it as the major impediment to rapid progress on Aquila; and the workers were tired of being caught amidst the increasingly frequent managerial squabbles.

Thus there was mass relief on the day in July when *Macs' Effort* was loaded onto the travel lift and dragged from the shop like the slab of loathsome meat it'd become in everyone's mind. After a day or two to install the rudder and propeller **(Illustration 21)**, the vessel was ready for launching.

A launching is traditionally a festive occasion, the culmination of months of exacting work, the celebration of a job well done, the christening of a newborn entity. As chance would have it, the high tide most favorable for launching came at night that July. Everyone who had the slightest relationship to *Macs' Effort*—the Knapp family and friends, the owners' families and friends, the boatyard workers, a banker or two—congregated in the dark on the dock at Hylebos Boat Haven. *Macs' Effort* hung in the travel lift, lit by floodlights.

Don tied a bottle of champagne to a rope hanging from the boat's bow. Bob's wife swung the bottle hard against the forefoot of the vessel. The bottle shattered and spilled champagne onto the dock, completing the time-honored libation to Poseidon. But due to the months of stress, anxiety, and animosity during construction, the occasion was subdued. As the travel lift lowered the *Effort* into the stygian blackness of the polluted waterway, it invoked to me the image of an unwed mother drowning her illegitimate baby on a starless night. Only later, when I understood the depth of the relationship between Don and Chuck Knapp, and how tragically *Macs' Effort* and Aquila figured in this relationship, did this image begin to make sense.

∽

Don Knapp was born in Dayton, Washington, on 30 May 1921, which put him at 57 years old when *Macs' Effort* was launched. His own father had come badly out of WWI, a little crazy and a heavy-duty practicing alcoholic, perhaps as the result of mustard gas and a head trauma. Don's father was a jack of all trades: he fished, he worked the wheat fields, he logged, but above all he was a gifted woodworker. As a boy, Don worked at his father's side, and his first foray into boatbuilding was in his mid-teens, when the two of them built a 23-foot skiff together.

Don got his driver's license at age 14—a big lad, he lied that he was 16—and started driving log trucks. When he really was 16, he was working in a sawmill when a moving belt caught his shirttail, jerked him backwards over a wooden crate, and broke his back. It didn't slow him down long, because by 1938 he'd put in enough time with big northwestern construction companies like J.W. Bailey and MacDonald to obtain his journeyman's card in Carpenters Union Local 407. During the winter of 1939–40, he helped build a 56-foot fishing vessel. Just before WWII, he worked with his father constructing barracks and other buildings at Fort Lewis. His father was a foreman, and Don soon became a foreman himself. By this time he had a wife, Ginny, and a daughter.

In 1942, after the US entered the war, Don enlisted in the Navy. He served as a ship's carpenter on an Auxiliary Tug Rescue, a 165-foot seagoing tug. He re-broke his back during a typhoon on the way to Hawaii and was discharged in 1944. After Don underwent rehabilitation in Glenwood Springs, Colorado, he and his family returned to the Tacoma area, where he remained the rest of his life.

After the war, Don worked at various jobs. He opened a café in Parkland, next to the campus of Pacific Lutheran University, and was successful until the university instituted a dining hall and put him out of business. He sold Oldsmobiles for a couple of years at City Motors in Tacoma, but didn't like that well enough to stay with it. Around 1954, he went back to what he knew best and loved: carpentry. He built anything people would pay him for, but primarily cabinets. It is unclear where he learned this particular branch of woodworking—whether from his father, the Navy, or on his own—but he was exceptionally good at it.

By this time, the Knapps were living in Midland, 10 miles south of Tacoma. Don constructed a big cabinet shop, 2500 square feet, next to his house. One summer he took a job building a residence on a lake, so he built for himself a 15-foot, hard-chined plywood boat in order to enjoy the lake. Chuck was two years old at the time, and that was the first boat he remembered. He played with tools in the shop while Don worked; in essence, he learned carpentry by the Suzuki method One of his first accomplishments with hand tools was to saw a sawhorse in two.

Several years later, when Chuck was six, Don began a 19-foot planing hull with sawn oak frames, planked with marine mahogany plywood. This was for a family friend, Bud Wolreth, who for the next 20 would be the Knapps' mechanic on boat jobs. Done in spare time, this project took a year and a half, and Chuck helped throughout. Don stimulated his son's interest by asking his opinion on matters of construction and design, and let him start using a band saw when he was seven. Nowadays, this could get a parent slapped with a child-abuse charge.

From then on, Don and Chuck worked together after school and on weekends, whenever there was boat work to be done. And there was always a boat under construction. Soon after the 19 footer, they began a Monk-designed, 23-foot inboard cabin cruiser for another friend of the family. The friend soon backed out, but Don liked the boat so much that he went ahead with it. Partly it was a nostalgic undertaking, for this was much like the boat Don as a teenager had helped his own father build. Again a spare-time project, the cabin cruiser took father and son two years to complete. When it was done, they named it *Ginny Lee*, after wife and mother. They moored the boat on saltwater, and Chuck at age nine started learning to run it.

It was about this time that Don became interested in charter fishing. He was a fanatic sport fisherman, and every chance he got, he took the family to the town of Westport at Grays Harbor on the outer Washington coast. One year he placed second in a salmon derby. While sport fishing in Westport, he observed the burgeoning charter industry in the small port. He learned from charter captains that with a relatively small investment in a boat, an owner-skipper could make a year's wages in five months.

Don's interest was fueled, too, by stories of his grandfather. According to family legend, his grandfather had made a fortune in the charter-fishing business in Florida in the 1920s and 1930s. He'd taken out wealthy clients on a luxurious 65-foot Chris Craft, and with his earnings—according to the legend—had bought a hotel where many members of the Barnum and Bailey Circus spent their winters. The legend may have had some truth to it, for when the grandfather died, apparently murdered at sea, Don got a small inheritance. He cleverly locked this money away by buying two acres of land with it.

Around 1962, Don began looking for a way to obtain a boat at least 32 feet long that he could charter. He had meager resources, and the cheapest way to obtain a boat this size was to purchase a hull and finish the boat himself. For a year, he and Chuck scoured boatyards, attended government surplus auctions, and tramped the marinas looking at boats. Finally they found a brand new, 36-foot, sawn-frame, hard-chined, plywood gillnetter hull that the legendary John A. Martinolich, patriarch of the Martinolich boatbuilding empire, was willing to trade for Don's two acres of land. Don was an inveterate trader; for him this was a deal made in heaven.

For a month or two, the hull just sat in Don's shop while he and his son pondered it, taking measurements, making drawings on scraps of plywood of possible designs for the cabin profile and interior layout. During the same period, they visited boatyards and marinas for ideas. They scoured southern Puget Sound for engines, hardware, and electrical supplies at auctions and surplus outlets. They delved into Coast Guard regulations to ensure their construction would pass inspection. Three decades apart in age, they were classmates, soaking up boat knowledge like dry sponges.

Construction was slow. Don continued with cabinetmaking and other carpentry jobs to keep the family afloat. After a year and a half of borderline poverty and late hours working on the boat, Don realized he'd need financing to finish the boat in time for the next fishing season, so he took out a bank loan to allow him to work full time on the vessel. The Knapps launched the boat, *Ginny Lee II*, in July 1966 and ran her the 265 miles from Tacoma to Westport to fish out the remainder of that season. Chuck, then aged 13, started out as a deckhand with his father but after several weeks took a job on another boat.

"I wanted to work for somebody who knew his business, and my dad didn't yet," Chuck recalled. "That was his first year; he didn't really know how to fish, and he had a slow boat. I'd watch these other guys with faster boats who always seemed to be catching more fish, and I wanted to work for one of them. Part of it was that I just wanted to be out from under my dad's wing.

"That first season, I crewed on the *Patty Ann*, an almost-new 45 footer built by Jones-Goodell. I learned some from that guy. The following two years, I worked for Jack Simmons, a very intelligent, lively, no-bullshit, no-frills guy. His 55-footer was a great boat to work, and Jack was a natural teacher. I learned the basics of charter fishing from him."

The fishing season in Westport ran officially from April 15 to October 15, with the real meat of the season the five months from May 1 to September 30. Charter boats were busy in May and September, but absolutely plugged with clients from June through August. In the spring and fall, while school was underway, Chuck would deckhand on weekends. As soon as school was out, the whole family would move to Westport, where they kept a mobile home. Don and Chuck would be on the water nearly every day until Labor Day, sometimes 60 to 70 days in a row. After the season, they'd run *Ginny Lee II* back to Tacoma.

Whereas Don and Chuck lived for charter fishing, the rest of the family barely tolerated it. In naming boats after his wife, Don might have been trying to appease her, for Ginny hated charter fishing with a passion. Her idea of a good provider was someone who—unlike Don—plugged into an 8 to 5 job with benefits and brought home a steady paycheck. She argued against every one of Don's boats, from the first to the last.

"Mom was involved in our charter operations, well, kind of like a sea anchor," Chuck told me. "She brought us food when we worked late in the shop, and she made the annual trip from Tacoma to Westport. To be fair, Mom would bitch, argue with Dad, and undermine his confidence in what he was doing, but she would come along. She was bitchy but cooperative."

During this period when Don was running *Ginny Lee II*, he worked several winters in boatyards. He helped build heavy-framed, heavy-planked trollers at Marine View Boat and then wooden yachts for Norm Nordland. These spells in the yards were as close as Don ever got to any real training in boatbuilding. Mostly he taught himself.

As he gained experience fishing *Ginny Lee II*, Don slowly outgrew her. With a larger boat, he could expand his business from nearshore salmon charters to albacore, which required running 70 to 100 miles offshore. Over the winter of 1969–70, he and Chuck purchased a 60-foot Navy-surplus vessel, removed the cabin, and completely rebuilt the boat. This time, they knew better what they were doing—no more staring at a bare hull in fearful ignorance and drawing rough plans on scraps of plywood. They christened her *Ginny Lee III*, relaunched her in July 1970, and fished her together for the remainder of that season and the next.

"It was tough working for Dad," Chuck recalled, "but it was worth it to me at the time. *Ginny Lee III* hauled a lot of people. I had good people skills, worked very hard, and got a lot of tips. I made a ton of money working with him, and at the same time, I was increasingly involved in helping him make decisions about how and where to fish. Sometimes he'd go below for an afternoon nap, and I'd run the boat and the whole operation myself."

Chuck graduated from high school in 1971 with the intention of going on to college. He didn't really need college, if all he wanted to do was make a living, since he was already well advanced in a maritime career that could carry him the rest of his life. He'd helped build three non-trivial boats, had five years of experience on the water, and had begun his apprenticeship as a skipper. Further formal education was almost irrelevant to his future success, at least in financial terms. But he'd also graduated class president, and he had enough left in his savings after buying a 1967 Pontiac Firebird to cover his first year of college, and then some. So attend college he did, a year of it at Central Washington State in Ellensburg. If not a total disaster, it was at least not very productive.

"I dropped out after the first year," Chuck recalled. "I remember thinking: this place is full of wankers who don't know what they're doing, just killing time before they have to get serious about their lives. Unlike most of them, I was paying for my education myself."

The most useful thing to come from the year of college was that Chuck had time to study for his first skipper's license. After five and a half seasons as a crewman, he'd accumulated enough sea time to qualify for the most basic professional license, the OUPV (Operator Uninspected Passenger Vessels). In addition to a minimum of 360 days of documented experience on boats, this license requires that the candidate pass a rigorous written exam on navigation, rules of the road, mechanical systems, seamanship, lifesaving, firefighting, first aid, and radio operation.

Because the OUPV license permits its holder to carry no more than six paying passengers, it is commonly referred to as the "six-pack" license. There is a great deal of irony in this term. The license grants a captain as young as 18, who cannot legally buy a six pack of beer, life-and-death responsibility over six paying passengers and his crew on the sea, an environment notorious for killing humans.

Chuck passed the Coast Guard examination in the spring and obtained the license at age 19, just in time for the 1972 season. The license permitted him to operate a vessel up to 20 miles offshore between Port Angeles and Leadbetter Point, Washington. The vessel he got hired to run during his first season as skipper was *Drifter*, a 36 footer powered by twin gasoline engines. Thirteen years old, *Drifter* was one of the pioneer vessels in the Westport charter fishery, but it'd been neglected and was in bad mechanical condition. That season, the engines broke down frequently, pumps burned out, belts flew off, and the steering broke. Chuck learned fast how to make on-the-spot repairs under difficult conditions.

To make matters worse, *Drifter* had the bare minimum of navigational equipment, compass and depth sounder only—no radar, no loran—and that summer was one of the foggiest in recent memory. The fog didn't lift at all during a 15-day stretch in July. *Drifter*'s crew would lose sight of land as soon as they left the dock in the morning and wouldn't see land again until they were within a stone's throw of the breakwater, just outside the harbor, when they returned in the afternoon.

"It's a complicated thing to take one of those boats out, even when you can see," Chuck explained to me. "Westport is a bar port, and on a shift from outgoing to incoming tide, the bar can go from being fairly peaceful with a lump on it, to nasty, dangerous, and impassable. Fortunately, when I first started running *Drifter*, I made sure I knew compass courses out the buoy line, run times on the course, depth cross coordinates, buoy sounds, current speeds and directions, and so forth. By the time the pea-soup fog set in for weeks on end, I was very good at getting out and back. I never got lost. But trying to navigate and catch fish at the same time was intense. That first season as skipper, I was constantly nervous.

"On top of everything else, I spent at least a third of my time as a charter skipper keeping people from freaking out, individually and as a group. Many of the people I took fishing had never been on the ocean, let alone out of sight of land on rough water. It was absolutely vital to communicate to them exactly what was happening and what was expected of them. Some people got drunk on board, and I had to keep them from hurting themselves or anyone else. Also, many of my clients would arrive at Westport tired and hungry after driving all night. Some of them got horribly seasick, and with fear and uncertainty on top of that, they'd become stressed to the point where they lost all ability to cope reasonably. I had to help them deal with their crises so as not to have to take them back to shore."

A lot happened just after the 1972 season. Chuck went into partnership with his soon-to-be father-in-law to buy a charter office, Shamrock Charters. A charter office is a business that manages a group of charter boats, similarly to the way a real estate office sells houses for people. The office advertises the boats; makes reservations for them; interfaces with other branches of the travel industry such as package tours; handles the paperwork for clients; and rents fishing poles. For these services, the charter office receives a commission from each boat. A fisherman with a boat and a charter office not only makes money from his own boat, but siphons off money from other boats. A charter office can own boats as well, paying skippers a daily wage to run them. Shamrock Charters was a well-established business, centrally located, with five boats in operation and a loyal clientele.

That same year, Chuck encountered a guy selling a used 40-foot crash-boat hull cheap—only $3000—and spent some of his savings on it. A "crash boat" is a type of high-speed Navy vessel with a deep V forward and a double-planked hull, designed to rescue people from crashed airplanes in rough seas. The hull was sound, and at first Chuck intended simply to slap a couple of inexpensive engines in it and be on his way. Then he thought, hell, why not do a really nice job of it, replace the pilothouse and cabin with a modern design, update the equipment?

To rebuild the superstructure, he first had to tear out the old one, which he did the fall after the 1972 season. "It was very interesting," Chuck recalled. "I got to dissect a boat. Not the hull, but the cabin and deck structure. It took me a month. The boat was moored in Hylebos Waterway, and I worked by myself, tearing it apart and throwing it out on the floating dock. Once I almost sank the dock. Anyway, I got to see how A. M. Rambo Company in San Diego had built the thing. I learned which techniques were sound, and which ones lacked integrity. For example, in places where they used only nails, I could take the boat apart with a crowbar. Where they used screws and glue, I had to chop it apart. That made a big impression on me."

When the boat was gutted, Chuck moved it into the shop at Hylebos Boat Haven. He drew formal plans for a new cabin and interior, and once again he and Don worked together, putting in 12- to 16-hour days through the winter of 1972–73 in order to have the boat ready for the next season. During that period, Chuck also studied for his Ocean Operator's License, since he now had enough sea time to qualify, 720 days. This license would allow him to transport any number of passengers up to 100 miles offshore, on inspected vessels of up to 100 gross tons displacement. By June 1973, Chuck had launched the vessel, *Malibu*, and he had his license.

Malibu was a great success. Mechanically sound and comfortable to fish, it also proved to be the third-fastest boat among the 220-vessel Westport charter fleet. Chuck earned barrels of money with it that summer. *Malibu* made a favorable enough impression that a friend of the Knapps asked them to build him a boat for the next season. That was really the beginning of Knapp Boatbuilding. For the next four years, the Knapps followed a seasonal cycle whereby they fished in Westport from May through September and then built boats in Tacoma from October through April. They generally purchased new, stock fiberglass hulls, mostly Lynn Seynour designs made by Delta or Uniflite, and finished them from the hull up, as they did with *Macs' Effort*.

The first charter boat the Knapps built for someone else was the 43-foot *Cold Spaghetti*, launched in 1974 from the shop at Hylebos Boat Haven. By that time, Chuck at age 21, was an experienced boatbuilder with journeyman-level skills. He was capable of designing the cabin, interior, and mechanical layouts of a charter boat; of drawing plans of sufficient quality for Coast Guard certification; and of competently conducting any phase of construction except electrical installation.

Since their boatbuilding enterprise looked like it was going to be a success, after the 1974 fishing season Don Knapp leased the plot of land adjacent to the Boat Haven and incorporated Knapp Boatbuilding. He and Chuck then commenced to build the shop where I later worked. They didn't have time to screw around, since they had an order for a big boat for the next season. They started construction on the shop in November and finished it in January. Compared to boatbuilding, Chuck found standard carpentry to be "simple and unimaginative."

The first boat built in the new shop was 50-foot *CharDan*. For the first time in their long partnership, the Knapps hired a construction crew, four carpenters. Don, now 53 years old, removed himself from the actual construction and confined his activities to ordering and business. Chuck functioned as foreman, overseeing and participating in the work. They finished *CharDan* in five months, launching it in June 1975.

The following winter, Knapp Boatbuilding undertook two relatively minor jobs with a construction crew of two men. One was remodeling a 34-foot power-cruising yacht. The other involved replacing the mechanical and electrical systems, finishing the interior, and handling the commissioning of a 40-foot charter boat, *Tabby*.

"That winter I was not very busy," Chuck recalled, "so I took the opportunity to study. The books I used most were *Boatbuilding Manual* by Robert Stewart and *Skene's Elements of Yacht Design* by Francis Kinney. Both were excellent references that overlapped one another in subject matter. At that point I was hungry to read; I had already seen and done so much of what those books discussed. *Skene's Elements* was a challenge, though, because it included design formulas. I didn't understand many of the formulas until later, when I applied them."

In 1976, Chuck decided to sell *Malibu* without the name and to build a new *Malibu* on a 43-foot stock Delta hull, using the same basic design he'd used for *Cold Spaghetti* three years previously. Before the 1976 fishing season was over, he had a buyer lined up for the old *Malibu* and financing for the new one. Another fisherman wanted a boat identical to the one he intended to build, so the Knapps with the help of an expanded construction crew built two sister boats side-by-side, new *Malibu* and *Windsong*. They poured into these boats everything they'd learned in 10 years of charter fishing, making design improvements to the pilot station, sleeping accommodations, engine room, mechanical systems, and overall structural integrity. Chuck invented a combined bait tank and fish-cleaning station made of aluminum, a design that he was later able to sell.

Chuck fished the new *Malibu* in 1977 and had the best season of his career, making around $40,000 in five months. Then, with a prescience stockbrokers would kill for, he realized the charter fishing industry was about to go bust, and he got out in a hurry.

"In 1977, fishing got quiet toward the end of the season in a really unusual way," Chuck recalled. "It seemed like the fleet had gotten too big at that point; there were maybe 275 boats in Westport by then. Also, I had started feeling very nervous about the Boldt Decision. I decided to sell my boat and license, and I had a buyer within two weeks. I got absolute top dollar. I got more for my boat after using it one season than any 43 footer had ever sold for new in Westport. A year later, you could hardly give a boat away. I mean, I just totally caught that one."

The Boldt Decision was a highly controversial 1974 ruling by Federal Judge George Boldt that reaffirmed the historical treaty rights of Native Americans in Washington State to fish at their traditional sites. Not only that, but the judge's decision had some teeth to it. It allocated to the Natives, who comprised 1% percent of the state's population, 50% of the harvestable salmon running through their traditional waters—that is to say, most of coastal Washington.

The decision, of course, enraged non-Native fishermen and led to sometimes-violent protests, but those were nothing new. The State had been jailing Native fisherman for decades; Marlin Brando himself had come to support them in 1964 and was arrested on the Puyallup River. In 1977, the Boldt Decision was still grinding its way through the appeal process on its way to the Supreme Court, and the outcome was highly uncertain. What was certain was that if the ruling were upheld, it would lead to significantly reduced catch allocations for the increasingly large Westport charter fleet.

By the time Chuck quit charter fishing, Knapp Boatbuilding had gained a solid reputation for quality custom work. Even so, Chuck was thinking of quitting boatbuilding as well, for reasons that went deeper than the mere premonition of a failing fishery. Maybe he was just burning out.

Now aged 24, he'd spent exactly half the summers of his life on the water, and he'd worked under considerable pressure the previous five winters to finish boats in time for the fishing season. His marriage to his high-school sweetheart had blown up a couple years before, and though he never talked about it, that may have been a wound festering under the salve of incessant work. Or maybe he felt chained to his father, with whom he'd been working since childhood.

I suspect, however, that Chuck simply wanted to try something new, to explore the intellectual world beyond the small province of charter fishing and boatbuilding. At that time, he was becoming increasingly interested in what he described as "philosophical and metaphysical kinds of mystery." The summer we worked on Aquila, he set aside his boatbuilding books and plunged into the works of Gurdjieff, an Eastern mystic considered profound by some readers and wacko by others. He joined the Tacoma Astronomical Society and began working his way through the Messier objects, a list of must-see astronomical entities that amateur astronomers keep track of much in the same way amateur birdwatchers keep their life lists.

It seems contradictory, then, that at the same time he was trying to pull out of boatbuilding, Chuck would undertake to build *Macs' Effort*, the most complex job the Knapps had ever attempted. The *Effort* was a big boat designed to go very fast, and the faster the boat, the better balanced and lighter it must be. In boatbuilding parlance, the *Effort* had to be built to fine tolerances and have a high strength-to-weight ratio. Complicating these requirements was the length addition, which had to be visually and structurally seamless with the rest of the hull. As it turned out, *Macs' Effort* was to Knapp Boatbuilding what the iceberg was to *Titanic*.

It wasn't until much later that Chuck was willing to reflect on the ongoing disaster I stumbled into when I went to work at Knapp. "The first problem was," he said, "that the complexity of *Macs' Effort* exceeded our organizational capacity as a business—our capacity to bid the boat, manage it, track changes, and get it done on time and within budget. The business basically consisted of Dad and me. He was the more knowledgeable boatbuilder, but he was the one manning the office and I was the one thrashing around out on the job. If we'd had someone completely handling the office, and if Dad had been out there with his tools working full steam, I think it would have been a whole different story.

"The second problem was that we didn't analyze the bid very seriously—we just took a stab at it. In the end, we simply ran over, timewise and cost-wise. Toward the end, it looked like Dad was going to lose $30,000 on the boat. In those days, that was a lot of money. When it became clear what was happening, Dad just went into brain freeze; he couldn't deal with it. After we finished the boat, I had to spend the better part of two weeks going through all the changes and additions the owners had asked for, put a price on them, write them up, and present a bill. The owners didn't argue once I'd done that, and we recouped some of the money. But Dad was in an absolute funk.

"The third problem," Chuck went on, "was a really big fucking problem between Dad and me. In many ways, I was still a boneheaded kid trying to outdo his father, trying to have a leg up on him. Dad's attitude toward building was, if you know a technique works, if it's worked for other builders, if it's worked on your previous boats, then use it! Just get the thing done, make some money! But I was still exploring the process of boatbuilding; I wanted to optimize things. So, for example, instead of using sawn cabin and deck beams, which were perfectly adequate for that boat, I had the crew laminate the beams. I had them use screws in places we'd always gotten away with ring nails. Those things just flat cost more money, and I was a big enough butthead that when Dad objected, I'd up the emotional ante, and so would he."

"Like flipping him off that time?" I asked.

"Yeah, like that."

When I'd gone back to work at Knapp Boatbuilding and found it replaced by CLK Yacht Crafters, I'd framed a romantic picture in my mind about the changing of the guard, the march of generations, the aging warrior laying down his arms and ceding the kingdom to the loyal heir. But it wasn't quite like that, not quite like that at all.

"While I was struggling with the most complex boat I'd ever built," Chuck recalled, "what I really had in my mind to do was buy a sailboat and disappear into the sunset, just get away from it all. I was visiting boat brokers all around Puget Sound, and that was how I met Earl. Earl was working as a sales rep for several yacht builders and retail outlets. He knew sailboats and the sailboat market, and he wanted to learn to build them. We got to talking, and that was how we began on Aquila.

"Dad wasn't wild about the Aquila project; it smelled bad to him. He didn't like Earl at all, didn't trust him. So what I did, I just sort of commandeered the shop and the tools; I essentially bullied Dad out of the way and forced him into retirement."

The idea that Don could be forced to do anything against his will strained credulity beyond all rational limits. I couldn't imagine Don ceding his shop unless he really wanted Chuck to take it over, and I said so to Chuck.

"He was really reluctant. He was pissed about it. But he was also in shock from the way *Macs' Effort* was going, and he came to believe like the rest of us that Aquila could be financially very successful. And Aquila did have all the ingredients to be successful—except for a couple, that is," and here was a wry chuckle. "At the same time, we agreed to pay him well for leasing his shop and equipment. We insulated him from any liability in the new company. So what happened wasn't clearly one thing or another. On the one hand, he probably liked the idea of my taking over the business, making it successful, making him money on the deal. On the other hand, the timing was bad. He just wasn't ready to retire, not ready at all. He felt he still had boats to build."

All of us have had the experience of intruding, unexpected and unwanted, on a private situation. We enter a room where colleagues engaged in a vitriolic personal conversation suddenly fall silent. We drop in on a family whose dog has just been run over, or visit friends who, unbeknown to us, are enmeshed in divorce proceedings. If we encounter strangers undergoing a gut-wrenching experience, we might interpret their behavior under stress as their core personalities. This one is a jerk, that one is a bitch, the one over there a simpering fool. In these private dramas, we may not know what is wrong, but often we feel that something is wrong. What we are ignorant of in these circumstances is history, both proximate and ultimate.

When I began work at Knapp Boatbuilding, I intruded into a private situation on a grand scale. I saw only the surface of things, that Chuck and Don weren't getting along, that Don hated Earl. Don seemed ineffectual, inconsiderate, belligerent. Chuck seemed competent but harried. Earl seemed irrelevant. The owner seemed an asshole. My co-workers seemed stressed and unhappy. And in fact, everything was just as it seemed, but little of it had to do with the essences, the core personalities, of the people involved. What I didn't know at the time was history. When I finally learned that, it explained the impression I'd had at the launching of *Macs' Effort* of a funeral rather than a christening. Something had felt wrong; I just hadn't known what or why.

Chapter 10
Summer and fall of '78

The best starting technique is often spoilt by the tactics of opponents. ... It is usual for nearly all competitors in a race to want to start from the same position. This results in a fight for time, right of way, safe leeward position, etc., the outcome of which cannot be foretold.

—Juan Baader, *The Sailing Yacht*

Bates closed at the end of July for the annual summer break, which lasted through the first week in September. The break was a welcome relief from the unrelenting tedium of the school year. Trumbly used the time to work on *Windance* in his back yard. Students who didn't need to work took a vacation.

Willie Hartman and Armen Melkonian graduated from the program. Willie was having too much fun in Tacoma to return immediately to Kodiak. The fishing season would have been winding down anyway, so he went to work full-time at Martinac. Armen responded to a job advertisement in *WoodenBoat* magazine and went to work at a family-run yard in Maine that constructed traditional wooden boats. The rest of the class viewed Armen's choice with awe and skepticism; people "back East" were rumored to be unfriendly, and hidebound in their boatbuilding practices. Trumbly, though, was delighted with the prospect of a student spreading his teachings so far.

I planned to visit Alaska from the last week in August through the first week in September. The lease on my house was due to expire in September, and the landlord informed me that regrettably he needed to raise the rent. He claimed the City of Tacoma had raised taxes, but I suspect the real reason was the boat shed in the alley—either he wanted rent for that too or wanted me to move so he could get rid of it. Since I'd already moved my boat to Bates, I decided to find another place. Patrick Chapman went away to visit his parents, so I stayed at his apartment until I left for Alaska.

I had three weeks to work full time at CLK before I left. Chuck and Earl were anxious to make rapid progress on Aquila. Though neither Aquila's plug nor mold would ever touch water, both would nonetheless eat up money by the truckload with no hope for any return until a hull was cast and finished. It was necessary to get them done as quickly as possible.

Chuck, Earl, Griz, and I put in 10- to 12-hour days, six days a week. This schedule exceeded 40 hours a week, but none of us got overtime pay. Chuck and Earl worked for themselves; Griz was Earl's protégé and worked when Earl did, overtime or not. As a student, I somehow did not qualify for overtime pay, or so I was told. Not so with the Great Burba, who knew his rights as a journeyman. He refused to work overtime without appropriate compensation; therefore, CLK held him to straight time. Navy Chris, whose limited skills were in limited demand at this stage, worked part-time.

By the middle of August, the strongback and frames were complete. Burba and the others lined off the strongback and fastened the frames upside down on it by their cross-members (**Illustration 27**); they squared, plumbed, and braced each frame as they went along. This required only a day or two. The crew then cut the stem according to the pattern taken from the loft and fit it longitudinally into notches in the apices of the forward frames. Since the plug was not a real boat, the long keel member was made of two layers of 3/4-inch plywood nailed together. Its only function was to hold the frames firmly in place and provide a fastening surface for the strip planking along the midline of the plug.

While the rest of the crew was thus occupied, I worked on the transom, which was curved and reverse-raking. From Trumbly's lofting exercise, I'd learned how to deal with this kind of transom on the loft, and it was a simple matter to expand it and make a flat pattern.

Building the thing, however, proved to be a different proposition altogether. I'd watched a student build one at Bates and assumed I knew how to do it. It looked easy then, but Trumbly had been available for instant advice. Trumbly wasn't available now, and I had problems. The first time I cut the framing members, I got them wrong and had to start over. Then I had the devil of a time framing the transom to its true shape and covering it with plywood without distortion. In short, the transom was beyond my capabilities and took much longer than I expected.

One problem was that I'd been working long days and weekends for nearly a year, and I was simply burnt out. When I burn out, I invariably lose my appetite, and this happened then. I forced myself to eat to be able to continue working, but I was eating garbage. To save time, the CLK crew took quick meals in diners on the Flats a short drive from the boatyard. These diners were furnished in chrome and scarred vinyl as sterile as the Flats itself, and they specialized in grease. The one place that could marginally call itself a restaurant was Marilyn's, on the Pacific Highway. Lacking windows, Marilyn's was so dark inside you could scarcely see your food, which was probably a good thing. The house specialty was a massive chili dog smothered in cheese and onions, served with a mound of French fries. We joked that the fry cook kept a set of electroshock paddles under the counter to jump-start grease-jammed hearts.

My planned trip to Alaska, then, was as much a medical necessity as a vacation jaunt. I'd promised to complete the transom before I left and I succeeded, finishing it around midnight the night before my scheduled departure. I was so exhausted the next morning that I slept right through the alarm clock and awoke a half-hour before my plane departed Sea-Tac. It was an indication of how far Earl'd burrowed inside my head that all I could think to do was call him.

"Earl, I overslept and missed my flight. What should I do?" I asked, as though he could somehow summon back the airplane.

Earl laughed. "Hell, if I had a dollar for every flight I've missed, I could retire. Don't worry; your ticket is still good. All you need to do is call the airlines and reschedule for tomorrow."

I flew out the next day, wondering why I'd instinctively called Earl, of all people.

The first thing I did when I returned from Alaska was look for a place to live. It was with a sense of déjà vu, for I'd done the same thing almost exactly a year before. I soon found a basement apartment in an old, three-story house on Tacoma Avenue, a block from the first apartment I'd rented. With concrete walls and only two half-windows at street level, the apartment felt like a jail. It was too barren even for cockroaches. The rent was $50 a month; I remember this because a security deposit of the same amount became a contentious issue when I finally left Tacoma.

Instruction at Bates began one Monday morning as though there had been no interruption. Trumbly started class the way he did every

September, with a recitation of the rules. He spewed Trumblyisms port and starboard, with an occasional profanity thrown in for proper form. Now, though, I was among the Brahmans. We looked on with lofty indifference as the Trumbly Show astonished the new crop of Untouchables. I wasn't really indifferent to them; I knew exactly how they felt. As I had been a year before, however, they were strangers among a group that had worked together for more than 1200 hours. They were as infants, dumb in our common language.

Like a commander rescinding the furloughs of his troops in the face of an imminent enemy attack, Trumbly made it clear that if we old-timers thought we were going to have abundant time to work on our own projects in the coming year, we could think again. He now had 11 months to finish the T-38 before his retirement.

"We'll launch that sailboat by next August," he reiterated, "or we'll all die trying."

The prospect of finishing my own sailboat at Bates looked dismal.

When I returned to CLK, much progress was evident. The strip planking of Aquila's plug was more than three-quarters complete, and the sleek shape of the hull was much more apparent than with just the frames. As soon as he saw me, Burba let out a whistle and a whoop.

"Hoo-eee, it's about time you crawled back to face your mistakes. That piece of crap you built warped like a politician. Shows what happens when you let a boy do a man's job!"

I knew immediately that Burba was referring to the transom I'd built. I walked to the stern of the plug to examine it. It looked fine, but it also didn't look quite like my transom. Burba was so busy crowing that Chuck had to explain.

"When we mounted your transom," he said, "we found it was out of true. I don't really know why; it could have been the wood warping with the heat we've been having, or something to do with the glue drying. Actually, it wasn't very far off. But it was outside our limits of tolerance, so I had Burba do it over." Chuck kindly didn't mention the real possibility that I'd simply screwed up.

I crawled inside the plug to see how Burba'd built the thing. He'd constructed a massive, multi-tiered internal framework to hold the outer surface of the transom exactly in place until the strip planking could be nailed to the edges. It wasn't pretty, but by god it was sturdy; an earthquake wouldn't have shifted it a millimeter.

This wasn't at all how I'd learned to build a transom at Bates, but that was the whole problem. I'd prostrated myself before the goddess of dogma rather than considering the functional requirements of the thing. The transom on the plug wasn't an actual transom; it didn't have to be light and pretty. It would never feel the splash of waves, nor glide proudly into a marina. Its fate was to be hauled to the dump along with the rest of the plug as soon as the mold was done. Nonetheless, you didn't get along with Burba if you didn't give him a little of his own back.

"Admit it, Burba," I said, "you'd have built it the same way I did the first time around. You needed me to show you the way, that's all." And then, the ultimate insult: "Looks like a frigging house carpenter built it!"

"Don't worry," Burba said, grinning evilly. "Even though you cost 'em 40 hours they can't really afford, I don't think they'll fire you. No, they've got something much more unpleasant in mind."

It turned out Burba wasn't just blowing hot air. While I was gone, Chuck and Earl had decided to lower Aquila's sheer. This was easy to do on the lines drawing and loft simply by erasing and redrawing. Modifying the sheer of the plug itself, however, meant cutting away and re-fairing wood already in place. It was exactly the sort of medieval nightmare that lines and lofting were invented to prevent in the first place. Yet this was the task Chuck assigned me when I returned.

The sheer of a boat is nothing more than the angle where the hull meets the deck. When you view a boat from the side, you see the sheer as a graceful line—the sheer line—running the length of the boat. On most sailboats, the sheer line is slightly concave, curving gracefully downward from the bow to the middle, and then rising again toward the transom. On some sailboats and many powerboats, however, the sheer line is straight or even convex. If a boat has a convex sheer line, it is said to have "reverse sheer." Whatever its shape, the sheer line contributes heavily to the aesthetics of a boat. Even slight changes can significantly affect the overall appearance, and an unfair sheer line can ruin it.

When I'd made the frames for the plug, I'd cut their top ends at a precise height and angle determined from the loft. The reason for this was to allow us to continue the side planking of the plug inward onto the frame tops, forming a flange six inches wide around the entire sheer. This flange would be part of the mold and, in turn, be present on hulls cast from the mold. The flange would provide a means of eventually attaching the deck to the hull.

To lower the sheer on the plug, I first had to correct the loft and make new patterns for the tops of the frames. I then had to line up the new patterns on the upside-down frames, mark where to trim the latter, do the actual trimming with a saber saw, and re-fair the frame tops by hand with a block plane. Only when this was done could I proceed to re-install the sheer clamp.

This was straightforward work, but it was not easy. The plug was upside down on the strongback, with the sheer situated 2–3 feet from the floor. This required me to work sitting, kneeling, or lying on my back like an auto mechanic. My arms got so tired from working overhead that I could hardly lift them. I cursed Chuck and Earl, silently or sometimes aloud.

When Burba heard me muttering, he'd shout from high up on the plug, "How's it going down there in the dungeon?"

Indeed, during the weeks I labored on the sheer, I was the butt of the boatyard's jokes. Chuck and Earl would say with mock seriousness to the other workers, "You'd better do a good job, or we'll put you on punishment detail," casting their eyes downward to where I was groveling in the debris of construction.

Just as I was growing quite weary of this ribbing, fate kindly delivered a replacement. When I arrived at CLK one day halfway through the fall, I learned that Chuck was in the hospital. Everyone acted concerned, but when they talked about it, there were smirks, then snickering, and finally the whole crew broke out in hysterical laughter, tears running down their cheeks.

"You should've been here," said Earl. "Chuck's been having pains in his groin and testicles for the last few days, but he thought it was just a hernia and ignored it. Finally, yesterday evening, he got the cold sweats and was hurting so bad he could hardly stand, so we drove him straight to the emergency room.

"Of course, we couldn't let an opportunity like this go by, and we kidded him the whole way, you know, like 'You just don't have the balls for the job anymore, eh Chuck?' and 'Don't worry, we're groin to the hospital now.' Chuck couldn't help laughing, and every time he laughed, he almost fainted. He kept screaming, 'Just shut up, you bastards.'"

"The doctor checked him right away, and it turned out to be a twisted testicle. The little thing'd turned round and round, and cut off its own blood supply. It literally hung itself." And Earl laughed so hard at his new pun, he could scarcely stand up, "Aaaaaaah, ha, ha, ha . . ."

Griz took over the story. "So ... ha, ha ... so at that point, with the doctor and a couple of nurses standing around, everyone looking serious, discussing the ramifications of a half-castration, I sang that line from the Beatles song Yesterday, '... and now I'm only half the man I used to be,' and that did it, everyone went into uncontrollable hysterics, the doctor, the nurses, us, and Chuck, though Chuck was also screaming.

"Finally one of the nurses got herself under control and ejected us from the emergency room. We called Don and he soon showed up, all concerned. We told him that the situation was pretty balled up. He wasn't amused and made it clear he didn't want us around, so we left."

Chuck was back at work within a few days, though he hobbled painfully like a dog that's had too many fights.

"What I had was called torsion," he told me, "and it's supposed to be one of the most painful things a guy can get. I think it was caused by having overenthusiastic sex in a rough position. Anyway, the doctor didn't have to amputate—he just lashed my balls down so they'd never twist on me again."

For a while Chuck was the butt of endless jokes, but he took it well and people soon got tired of needling him.

꙰

Already in my second year of boatbuilding school, I was in a paradoxical situation. I was building a small sailboat, was helping build two big ones, had taken a course in celestial navigation, but did not yet know how to sail. Thus, when Chuck asked me if I wanted to fill in as a crewmember in a local sailing race aboard *White Squall*, a Cal 40 (40 footer) from the Tacoma Yacht Club, I jumped at the chance.

White Squall was owned and captained by a physician named Govner Teats, whom every called "Gov." In his seventies, Gov was semi-retired; he still practiced a bit of medicine, but his overriding passion was sailboat racing. A short, white-haired man with liver spots and a perpetual frown, Gov matched Captain Ahab in personality, with Gov's white whale being the trophy awarded annually to the top racing boat at the yacht club. Like Ahab, he was oblivious to any manifestation of weather or consideration for his crew, except as these affected his progress toward the finish line. He didn't race to enjoy a nice day on the water; every time *White Squall* crossed the starting line, his sole goal was to win.

The only complication in Gov's single-minded pursuit of the trophy was that he couldn't race *White Squall* alone but needed a crew of five or six besides himself. His success in racing depended entirely upon the experience of his crew, and experienced crewmen were difficult to find and keep. People left town, quit, or moved to other boats. On any given weekend, a crewmember might be absent due to a personal conflict or illness. Thus Gov continually tried out new crewmen. If they had experience and there was a vacancy, he invited them to join the regular crew. If they looked promising, he added them to his list of backups.

Chuck and Earl were regular crew on *White Squall*. Our initial connection to Gov Teats was Earl, who'd been an avid sailor for 20 years. Chuck had started on *White Squall* the previous season and thus already had a lot more experience than I. The only other regular crewmembers I remember were John, a retired air-force officer in his fifties who owned an electronics firm; John's daughter Grace, about my age; and Paul, a science journalist in his thirties.

Given Chuck's prior experience on the water, it was no surprise that Gov had snapped him up as regular crew. He knew the sea, the rules of the road, coastal navigation, and everything else it takes to handle a boat safely in nearshore waters. Nonetheless, all this did not mean Chuck was yet a competent sailor, for powerboating is as different from sailing as motorcycling is from surfing. The only knowledge one needs to move a powerboat from point A to point B is how to start the motor, engage the gears, and apply the throttle. Anyone who can drive an automobile thus has the basic skills necessary to drive a powerboat, and the plethora of drunken fools in runabouts that plagues US waterways every summer bears this out.

In contrast, even the most experienced powerboater finds a sailboat to be a complicated and terrifying contraption the first time he boards one. The profusion of sails, lines, winches, pulleys, and cleats is daunting, to say nothing of manipulating these paraphernalia to move the vessel. One does not simply point a sailboat toward a destination and expect to get there directly, for the direction and velocity of the wind fundamentally limit the speed and direction of travel. Nor can one learn to sail from a book. Handling the complex gear must become as reflexive to the sailor as applying the brake is to an automobile driver. If a driver suddenly veering toward a brick wall has to think about depressing the brake, he's already a couple seconds too late.

My first race was a Saturday afternoon event around a few buoys on Commencement Bay. A sailboat race does not start like a foot race, where the runners hunch over at the starting line and then sprint away at the sound of the gun. This might be the rational way to do it—have all competing boats anchor on a line. At the start signal, all boats could weigh anchor and get under sail. But this is not how the sport evolved. Instead, there is an imaginary starting line between two buoys. All the boats jockey for position upwind of this line and attempt to cross it as soon as possible after the start signal.

A successful start can be a major factor in winning a sailboat race. A boat that gains the lead at the start and maintains this lead will win the race. Because some boats are faster than others, and some crews more skilled, being first across the starting line does not guarantee a win, but if all else is equal it's a big advantage.

The optimal start is to cross the line just after the start signal, at the top speed possible under the prevailing conditions. In theory, this is not difficult for an experienced captain. The captain's watch is synchronized with the race clock, and he knows the start time to the second. He can estimate how long it will take to get from a particular position to the start line. As he nears the line, he can slow his speed if necessary by spilling air from the sails, so that he will cross just after the gun. The only complication is that fifteen or twenty other captains are trying to do exactly the same thing across a line that is too short to accommodate all of them simultaneously.

Therefore, captains employ bluff and treachery. For example, there is a rule that a boat on starboard tack—one headed so that the wind is coming from the right—has the right-of-way over a boat on port tack. A captain approaching the line on starboard tack will thus have an advantage over all boats on port tack, for the latter must get out of his way. Usually they do, but sometimes a captain on the port tack will pretend not to see a boat on starboard tack. This, of course, is illegal, but the captain who has the right-of-way will wonder, "Does this guy really not see me, or is he just bluffing?" If he is obstinate, the captain with right-of-way will call the bluff and risk damaging his boat. If not, he will veer off and lodge a complaint after the race.

I'd expected a sailboat race would be a relaxed, congenial affair. Certainly the start was a festive scene. Boats with hulls and sails ranging the whole color spectrum zigzagged over the whitecaps like a feeding frenzy of gulls, bright against the sea. Masts clanked and sails fluttered and popped as the boats tacked amongst one another, their flags flapping in the breeze. But it was also a rabid orgy of testosterone and adrenaline. Gentlemen sailors screamed at one another and swore at their crews. Yachts played chicken with other yachts, each of them worth anywhere from tens to hundreds of thousands of dollars apiece. Two collisions happened near the starting line, sickening thuds with a wrenching of intertwined rigging, obscenities and accusations flying.

The rest of the race wasn't much different. Gov allowed us no time to admire Tacoma from the water, to grow pensive over the hazy blue of Commencement Bay in the autumn sun. He demanded we focus our entire attention on *White Squall*. If he saw a crewman's attention wandering, he'd shout, "Watch those telltales," meaning the streamers sewn to the sails to indicate airflow, or "Tell me the instant that boat tacks," meaning one of the competition. If he couldn't think of anything else, he'd shout, "Wake up, there! Pay attention!"

Gov had designated John to be first mate, and the two of them constantly conferred over the best course to avoid slack winds in the lee of land. They planned moves like a chess strategy to come upwind of other boats, thus putting them in our wind shadow and decreasing their speed. They bluffed in rounding buoys, where collisions sometimes happened just as at the start.

In short, sailboat racing wasn't at all what I'd expected, and I decided on that very first race I didn't much like it. I wasn't learning to sail; all I was doing was hauling lines and turning winches. I could care less how fast we got to the finish line, or indeed whether we got there at all. I'd much have preferred meandering about the bay, fishing and exploring small islands, but I didn't know a boat owner who wanted to do those things.

Though I was only a backup, I participated in most of the races *White Squall* entered in the 1978–79 season, and none of them was what I'd call pleasant. Though winter air and sea-surface temperatures in Puget Sound are not as cold as in Alaska, they are cold enough to be dangerous. Some races lasted two days and the intervening night. A helmsman and two crewmembers at a time did night watches on staggered four-hour shifts, and we grew numb with cold from the constant wind and blown spray.

The ability to warm up for a few minutes during watches, or even between watches, would have made night sailing much more bearable. It would also have increased our ability to respond to emergencies, for few people function well in the early stages of hypothermia. But Gov had a Spartan attitude toward racing; although *White Squall* had a heater in the cabin, he refused to light it.

A sailboat demands constant vigilance even under normal circumstances. On every tack, the boom sweeps above the deck from one side to the other, and if you don't duck, it will brain you. In racing, the boat is constantly pushed to the margin of safety, and the only thing you can really expect is that the unexpected will eventually occur. This lesson was driven home once when we were sailing *White Squall* at dusk on an overnight winter race. There was a steady breeze, and we had a spinnaker up. A spinnaker is a large, colorful, parachute-like sail that billows out in front of the vessel. It is used in conjunction with the other sails to make maximum speed in a light to moderate trailing wind. As we were sailing along fast, a strong gust blew in out of nowhere at an angle to the prevailing wind. This shifted the spinnaker off to one side and then tore it like tissue. *White Squall* broached, or turned sideways to the wind; the wind caught the other sails sideways and knocked her over. The mast and mainsail hit the water, and we found ourselves hanging on to the now-vertical deck, tangled in lines. The capsize took no more than 10 seconds.

Amid the chaos, Gov and the experienced crewmen knew exactly what to do, without discussion or redundancy. They released the halyard to drop the mainsail as *White Squall*'s ballasted keel righted her, got the boat pointed into the wind, untangled the snarl of rigging, and coaxed her underway again. Inexperienced, I was less than useless; I was a liability. The best I could do was hold on, stay out of the way, and turn winches as directed.

I finally formulated a theory why people race sailboats in the winter, especially in northern climes: their workaday lives are so tedious and boring that they subject themselves to a decidedly unpleasant experience every other weekend so that their daily routines seem pleasant by comparison. In other words, winter sailing is like hitting yourself in the head with a hammer because it feels good when it stops. Of course, this logic could be applied to a lot of things, like winter mountaineering and running marathons.

Okay, I can accept that competitive sailing was just not my cup of tea; different strokes for different folks, and all that. The thing was, no one on *White Squall* appeared to be having buckets of fun. There was little camaraderie and no esprit de corps. Gov and his regular crew weren't friendly; they scarcely talked to me, other than as a piece of machinery for turning a winch. They weren't free with knowledge or patient in teaching. I could perhaps understand their attitude toward me—after all, I was only a voc-school student and a rank beginner at sailing—but they were downright unpleasant to Earl, which was a mystery.

Earl should have fit in aboard *White Squall*. He claimed to be a businessman of some standing; if he wasn't one, he at least had a great line of bullshit. He was conversant in professional sports, the predominant topic among the yachting clique. Earl claimed to have crewed in grueling SORC races off Florida and to have raced in southern California and Hawaii, and I later had some confirmation that this was true. Yet on *White Squall*, he could do nothing right. When he offered a suggestion, Gov and John ignored him. They frequently corrected his actions, as though he were a novice. Earl responded by growing increasingly surly as a race progressed; he'd sit in the cockpit with an angry scowl on his face, communicating in grunts.

Toward the end of the season, Earl irrevocably damaged his standing with Gov through an untoward quip. One of the crew commented he wished we'd done better in our race.

"We did the best we could, given what we had to work with," said Earl.

"What do you mean?"

"Well, a boat with both Teats and Dick has got to be a little slower than the rest."

It was a clever joke that even I thought was funny, but Gov overheard it and was deeply offended.

Chuck, Earl, and I stayed on *White Squall* as long as we could. My only goal was to learn to sail, but for Chuck and Earl sailboat racing was a matter of professional expediency. To know how best to build the Aquila line, to become familiar with hardware and rigging, and to eventually demonstrate the boats, Chuck needed to be knowledgeable about racing. Earl, who would be in charge of marketing Aquila, needed to make and maintain contacts. Captains in the yacht clubs around Puget Sound constituted his closest reservoir of potential customers, and racing was the best way to get chummy with them.

⌐

The Tacoma Yacht Club, or TYC, comprised (it is still there, by the way) a modern clubhouse and marina at the edge of Point Defiance Park. More than simply a haven for boat owners to tie up yachts, TYC was a social institution. Joining the club required nomination by two members, approval by the Membership Committee and the Board of Trustees, a credit check, and an initiation fee—all of which served to keep out the riffraff, in whatever way riffraff was defined. The club building housed a bar and restaurant where members could entertain friends or business associates, and a hall for catered receptions such as for weddings and birthdays. The club sponsored banquets, yacht parades, club cruises, and regattas to help its members fill their lives. Finally, TYC provided a venue for the alpha males among its members to exercise aggression in a socially acceptable manner through racing.

TYC was a private club, not open to the public but not entirely closed either. Unlike a golf or tennis club, its focal sport depended on hoi polloi like me for crew members, who participated directly alongside club members—analogous to a group of CEOs regularly inviting their caddies to join them as players in rounds of golf.

Whenever I think of TYC, one incident in particular comes to mind. Late one winter afternoon after a race, all the captains and their crews congregated in the clubhouse to warm up and celebrate. It was near Christmas, and so the mood was unusually festive. A long line of singing, drinking, shouting revelers, each holding the waist of the person in front, snaked its way among the tables, from one room to another. I was hunkered up to the bar, separate from the rest of *White Squall*'s crew. Two stools away sat a woman about my age. There was nothing remarkable about her. She wasn't beautiful, but she wasn't bad looking either. She was neither very tall nor very short. Her hair was dishwater-blond, neither glossy nor dirty. She seemed a little worn down.

"How you doin'? What brings you here?" she said when she saw me glancing her way.

"I'm a crew member on *White Squall*," I said. "Other than that, I'm a student in the boatbuilding program at Bates. How about yourself?"

"I'm a nurse," she said, "and I hate sailing. What I'm doing here is, I'm trying to catch that." And she pointed at a young, short, mousy guy who was part of the snake line. He had shaggy brown hair, a caterpillar moustache, and small eyes set too close together. "He's a doctor. He hasn't met me yet, but he's about to. You take care!"

With that, she hopped off her stool, just as the line broke momentarily behind the doctor. She slipped into the line, put one hand on the doctor's waist, and took a long gulp of her drink with the other. As the line shuffle-stepped away, she flashed a brief, grim smile back toward me, as if to say, "I hate this, but it's what I've got to do."

"All the best," I said, raising my glass as the drunken, idiotic snake dragged her away, hard on her prey (double entendre noted).

I was amazed at how frank she was to a stranger and how much we'd communicated in a few seconds. We'd each immediately recognized that the other didn't belong at TYC and that neither of us felt at home there.

Ironically, if she hooked the doctor, she might be spending a lot more time at TYC than she wanted. I never saw her again.

Chapter 11
Petrochemical hell

Sometimes there will be some difficulty in removing the mold from the plug (as well as removing a completed part from the mold). This problem is usually solved by a couple of solid raps with a mallet, or water or compressed air injected between the pieces which will usually release the sticky parts.

—Ken Hankinson, *How to Fiberglass Boats*

As soon as Aquila's plug was completely planked, Burba left CLK Yacht Crafters. There would be no further need for journeyman carpenters until it was time to finish a hull. The next stages involved sanding the plug smooth, laying up the fiberglass mold over the plug, polishing the mold, and then casting a hull. To provide expertise for these tasks, CLK hired a fiberglass specialist named Lyall.

Lyall was in his late forties to early fifties and had been working in the fiberglass industry since the 1950s. Like Burba, he had an ego the size of a battleship. He was a craftsman to the core, proud of his abilities and more than willing to enumerate them. Lyall looked healthy enough; he had a full head of brown hair with a speckling of gray in it, and he wasn't emaciated. Actually, he was a medical miracle, for he resolutely refused to wear a breathing mask while either laying up or grinding fiberglass. He sucked those little glass needles right in, soothed his lungs with full breaths of whatever organic vapors came his way, and then cauterized the wounds by smoking every chance he got.

"I've been fiberglassing my whole life without a mask, and I'm not dead yet, so it must not be all that bad for you," Lyall explained in a fit of rationalization his first day on the job, when someone made the mistake of asking him if he needed to borrow a mask.

Lyall's expertise was not simply the manual ability to paste together sheets of fiberglass with liquid resin—any chimpanzee can do that—nor was he a boatbuilder. CLK hired Lyall for what was in his brain, the instinctive knowledge of a material that comes only with long experience. Fiberglass is used to fabricate many other objects than boats, such as large-diameter pipes and septic tanks, and Lyall had done them all. He was an oracle on fiberglassing.

Much work remained before we could lay up the mold over the plug. We filled gaps between the strip planks with a plastic compound that dried to a firm but sandable consistency and planed down high spots resulting from slight irregularities in plank thickness. Furthermore, in making a mold from a plug, absolutely the worst thing that can happen is for the two to stick together. If this happens, the only recourse is to discard both and start over. Therefore Lyall directed the coating of the entire plug with white resin mixed with "microballoons," a powder of microscopic plastic spheres. This both sealed the pores of the wood, thus reducing the chance the mold would stick, and provided a surface of uniform hardness suitable for sanding. In sanding, it is very easy to leave flat spots, so we experimented until we found the optimal length and flexibility of sanding board to avoid this problem. We then sanded the plug and coated it some more, sanded and coated, sanded and coated **(Illustration 28a)**.

When the plug was as fair and smooth as a baby's bottom, we applied a "parting agent," a special wax used to prevent the mold from sticking. To be sure of not missing an iota of the surface, we rubbed on several coats with cloths, just as one might polish an automobile. Then, as an added precaution, Lyall sprayed a second parting agent, polyvinyl alcohol, over the wax.

With the plug hopefully as unsticky as a Teflon-coated frying pan, Lyall himself sprayed on the first layer of the mold—a black, hard-curing resin called "tooling gel coat," which would allow us to later polish the inside of the mold. Because of its hardness, this gel coat is very brittle. If it is applied in too thick a layer, it can crack with flexion of the mold; if in too thin a layer, it will not provide enough material for polishing. The correct thickness of this first layer is crucial, which is why Lyall applied it himself. Through experience, he knew how to spray uniformly and could estimate the layer's thickness. Even so, when he sprayed the plug, he also sprayed several waxed test panels. When the gel coat had hardened on the panels, he removed this layer and measured its thickness with a micrometer.

Once the tooling gel coat was on, laying up the mold involved applying layer after layer of resin-soaked mat and roving over the surface of the plug. It was winter by the time we got to this stage, and nighttime temperatures in the shop were sometimes close to freezing. Since cold resin does not cure well, and since the shop was too large and drafty to heat in its entirety, we constructed a makeshift tent of translucent plastic around the plug and installed a forced-air furnace and thermostat. Even so, the temperature inside the tent fluctuated and was often below room temperature. This called for Lyall's expertise to estimate the proper amount of catalyst to add to the resin at any given temperature to allow it to cure in timely manner—neither too quickly nor too slowly. Since the strength of a fiberglass laminate depends on the proper ratio of resin to fiberglass, he also kept a sharp eye on how much resin we used.

All of us worked inside the tent, where conditions were a wood purist's nightmare and a glue-sniffer's paradise. We looked like aliens in a high-security isolation chamber at the Roswell UFO site. Our black breathing masks were like the mouthparts of Andromedan insects. Stiffened and glossy with blotches of hardened resin, everything we wore—boots, coveralls, long-sleeved shirts, and watch caps—had the appearance of diseased exoskeleton. We swarmed on scaffolding around the plug like worker creatures attending a giant, pupal queen.

Laying up the mold was similar to laying up the fiberglass dinghies I'd worked on, except that we included a layer of 3/4-inch-thick balsa wood **(Illustration 29)**. The balsa came in flexible sheets of 1-inch squares held together by fiberglass fabric. This balsa core thickened and stiffened the mold with a minimal increase in weight.

The mold was also amplified to a Titan scale compared to the dinghies. We cut yards upon yards of mat and roving, as though we were sewing a dress for the Statue of Liberty. We drew resin from 55-gallon drums into gallon cans, stirred in catalyst by the tablespoon, and literally poured the resin onto the mat or roving, spreading it with paint rollers. We removed air bubbles and excess resin from the laminate with grooved metal rollers and squeegees a foot long. We used acetone by the gallon to clean tools, and we practically bathed in the stuff to clean ourselves. Because acetone is highly flammable, we had to keep it in the cold shop, away from the furnace in the tent. Like alcohol, acetone cools a surface as it evaporates, and cleaning up with cold acetone numbed our hands as effectively as a glacial stream.

Anyone who works with fiberglass tacitly accepts the slogan "Better living through chemistry," but at CLK we carried it a step further. At some point in that bleak winter of cold acetone, we took to fortifying ourselves during lunches and coffee breaks from gallon jugs of cheap red wine we kept in a lunchroom cabinet, hidden well away from Don's sharp eyes. The wine bolstered our spirits so well that we soon added port to our chemical repertoire.

This was not a matter of disgruntled workers medicating themselves in defiance of management. No, management had their snouts right there in the trough as well. Among us, only Lyall declined to partake, but he was already perpetually high from organic fumes. I remember thinking at the time, "My god! What has American Industry come to?" Such musings, however, distracted me not a whit from improving my own life through chemistry.

If Don'd had the slightest inkling what was occurring in the lunchroom, there'd have been hell to pay. Though he was no longer the boss, he spent almost every day at the boatyard, which he still owned. He'd have found a way to fire everyone he could and would've slapped some sense into Chuck and Earl—he was tough enough to do it. His objection wouldn't have been our indulgence per se, but only that we indulged on company time. Don felt in his bones what American Industry was coming to, and he didn't like it one bit.

Curiously, relieved of the stress of *Macs' Effort* and the yard itself, Don proved not to have the fire-breathing, hair-trigger personality I'd initially attributed to him. Inside a shell that in times of anger was frightening to behold was trapped a considerate, soft-spoken, unexpectedly gentle man. This I learned in the unusual circumstance of hauling trash.

Through its routine operations, the boatyard generated a tremendous amount of trash, which we deposited in containers behind the shop. About once a month, Don hauled the accumulated trash to a dump in a big, flatbed truck with sidebars. The dump was an hour's drive from the yard, and with loading and unloading, a round trip required half a day

One Saturday, Chuck asked me if I would like to help Don with a dump run the following day. I was still deathly afraid of Don and tried to beg off. It seemed an outrageous request, like asking a mouse to take a drive with a cobra.

"You don't understand," Chuck said. "Dad particularly requested you and Griz to help him tomorrow. The thing is, dump runs are a ritual for him, and he won't take just anyone. Strange as it seems, it's an honor. Dad always follows the same routine. He loads the truck early and finishes unloading at the dump by mid-morning. On the way back, he treats whomever is with him to a big breakfast at his favorite roadside cafe. Asking you on a dump run means he likes you enough to eat Sunday breakfast with you."

I could hardly refuse a pitch like this, but I was still skeptical. The next morning was overcast but not raining. Don, Griz, and I together loaded the big truck. Don was as relaxed as I'd ever seen him. As he double-clutched the old truck on the way to the dump, he reminisced about his experiences as a carpenter, a charter fisherman, a boatbuilder.

The dump sat like a vast zit in a clearing in the woods. We unloaded our refuse to the cacophony of thousands of gulls circling above, their numerous cries melding into a single, maddening din. Many of the gulls landed what they considered a safe distance from us, ringing us like a stoning mob. This was their dump.

On the return trip, when we sat down to a sumptuous breakfast, Don seemed almost human. By the time we returned to Tacoma, he felt like a friend, and I never again feared him. From that day on, I looked forward to dump runs with Don.

⌐

To facilitate removal of the mold from the plug, and eventually hulls from the mold, we constructed the mold in two parts. This we did by attaching a temporary partition of plywood about eight inches high running vertically along the entire midline of the plug. We then laminated the first several layers of the mold over half of the plug, running the laminate up onto one side of this vertical partition. When the fiberglass had cured, we removed the partition, leaving a vertical flange on the first half of the mold. We waxed this flange and then laminated the first several layers of the other half of the mold, running it up onto the fiberglass flange on the first half. These abutting midline flanges would allow us to separate the mold from the plug in two halves, which we could then bolt together again through the flanges in order to to cast a hull.

When its walls had been built up to sufficient thickness, the mold was like a 54-foot-long bathtub weighing several tons, which presented two problems: without stiffening and support, it would tend to flex, and its weight would make it difficult to maneuver. To address the first problem, we fiberglassed longitudinal stiffeners—heavy cardboard tubes cut in half—onto the outer surface of the mold. We also attached five wide, transverse frames evenly spaced on the outside and surrounding the mold on the bottom and sides, each frame made of several layers of 3/4-inch plywood. The frames had flat outer edges that allowed the mold either to sit upright, or be tilted so that one side or the other lay nearly horizontal, making it easier to lay up fiberglass on the inside. Like the mold itself, the frames could be unbolted at the midline, so that the mold halves could be separated.

In December, the crew pulled the plug from the mold **(Illustration 28b)**. This occurred while I was out of town for the holidays. When I returned, the plug had been sawed into sections and deposited in the vacant lot next to the shop, the strongback had been disassembled, and the mold sat upright in their place. Pulling the plug had gone flawlessly, with only the usual difficulties inherent in prying apart two massive, rigid objects held tightly together by the suction of a microscopically thin layer of wax. Griz later told me that "popping the mold" had been like trying to take off a wet T-shirt without moving.

One last but important task remained before we could lay up a hull for Aquila inside the mold. We'd sanded the plug smooth, and so the inside of the mold was also smooth. But fiberglass hulls are not merely smooth; they have a polished surface. To impart this finish to a hull, we thus needed to polish the inside of the mold. This we did by laboriously hand-buffing it with a series of increasingly fine polishing compounds, and finally with jeweler's rouge. It was like polishing a giant telescope mirror, and like telescope makers we were fastidious. To avoid extraneous dirt that might leave scratches, we wore clean clothes and coveralls and worked in stockinged feet. Between changes in polishing compounds, we vacuumed and then wiped clean the entire inside surface of the mold.

With the mold finally completed, we could conceivably have the first hull within six weeks and finish a boat before the next racing season. Unfortunately, appearances were misleading. Though I had no inkling of it at the time, CLK was starting a rapid spiral down a financial toilet: the company was the fiscal equivalent of a movie set—a nice-looking storefront facade propped up by two-by-fours in back. Only later and with the wisdom of hindsight did certain incidents begin to make sense.

For example, a month or two previously, Chuck and Earl had left the yard one day for an appointment with a banker to obtain a $40,000 loan. This in itself was not an indication of financial difficulties, because most starting businesses borrow money. They got the loan, but Chuck made some cryptic remarks afterward that later proved to be apocalyptic.

"I let Earl do all the talking, and you should have heard him," Chuck told me. "He hinted at the fortune his mother had made in real estate; he played the guy like a violin. Earl was brilliant! By the end of our meeting, the banker was fawning all over him and couldn't wait to give us money; he was practically stuffing it in our pockets as we left.

"Oh, boy, if only he knew," Chuck concluded enigmatically, shaking his head with disbelief. Chuck was generally closemouthed about the inner workings of the boatyard; I knew it would be difficult to get any more information out of him, so I didn't pry.

Around the time we finished the mold, another incident occurred that also suggested financial difficulties. Everyone at the yard knew about my 14-foot sailboat. Some of my co-workers had given me rides home the previous summer and had seen it in the shed alongside my house. One day, Earl asked me how it was coming along. I told him I'd been working on the boat slowly but steadily at Bates throughout the fall and had completed the hull to the point of having applied the first coat of white primer paint.

"Well," he said, "I've got a business proposition for you. Why don't we take a mold from your hull? It's a nice-looking design, and maybe we can produce a bunch of fiberglass hulls fast. The yard can make some quick money, and you'll get a royalty on every hull we sell."

I thought about it. I didn't want to take time away from working on the boat. Also, knowing how much time it had taken to make and finish the smaller, simpler sailing dinghies for Roberts, I was skeptical that the project would be a source of much quick money. I told Earl my reservations.

"Listen, if you're willing, we'll bring your boat right down here and I'll have Griz lay up the mold. You won't have to do a thing. You can be involved in finishing the boats, or not, as you choose. If you work on them, you'll make an hourly wage. Otherwise, I'll get someone else to do them. Either way, you'll get the royalty, 10 percent of the sale price per boat. How does that sound?"

Earl was probing the same weakness in me that he'd found in the banker—greed. It was effective. A completely rigged boat this size might sell for $1500 or more, which would mean at least $150 a boat for me even if I did nothing. I liked the idea of making money by doing nothing, but I still wasn't completely convinced.

"There's something else, too," Earl said. "You may be a little touchy about this, but I think it's a hell of a marketing gimmick. You know that the official designation of a production boat is often the designer's name and the length. Your boat would be the Dick 14, but you know people are going to twist it around and call it the Fourteen Foot Dick."

I was more than a little tired of Earl's humor, and I began to suspect it was all a setup so he could make his little joke. He saw it on my face.

"I'm not joking," he continued. "Every product needs a catch in order to sell. Think about it. You know the yachting crowd; a lot of 'em are arrogant assholes. If they were in the market for a small sailboat, they'd buy this one just to be able to laugh and say they had a 14-Foot Dick. Sure, we could give it a touchy-feely name like the Surfbird 14 or the Sailfish 14, but a hundred other models have touchy-feely names. The 14-Foot Dick will sell."

I thought maybe Earl was a little insane, but greed outweighed my better judgment and pride, and I decided to see if he was serious. He could call it anything he wanted as long as I got my $150 per boat.

The following week, Griz and I hauled my boat from Bates to CLK in the ancient company pickup truck, and Griz immediately began on the mold. Since the boat was small, a single-piece mold was feasible. Griz supported the hull upside down on sawhorses, cleaned the surface well with acetone, and liberally waxed it to prevent sticking. He sprayed on the tooling gel coat. After that cured, he could begin laminating the mold.

Some days later, I came to work expecting to see the finished mold removed from the hull.

"It didn't work," Griz said as soon as he saw me.

"What do you mean, it didn't work?" I asked. "How could it not work?"

"Well, what happened was, the mold stuck to your boat like it was glued on, and I had a hell of a time getting it off. I spent all morning pounding on it, and it seemed like it was stuck for good. I thought both your boat and the mold were ruined. Finally I was able to cut away the mold in pieces and remove it, but even then it pulled some chunks out of your hull."

With a sinking feeling, I ran up the stairs up to the second floor, where my boat was. I'd immediately visualized gaping holes and was relieved to find that only the outer layer of the plywood had pulled away in a few small spots. I could easily fix them with filler.

"Why do you think it stuck? We did the same thing to Aquila's plug, and the mold didn't stick," I said.

"Well, when you think about it, we didn't do exactly the same thing. Your boat has a coat of primer paint on it, and the paint is oil-based. The oil that soaked into the wood probably dissolved the wax, which resulted in sticking. I actually thought about this beforehand, which is why I cleaned the hull several times with acetone, but obviously it wasn't enough to remove all the oil."

I was so relieved the hull was intact that I didn't regret too much my loss of royalties. Nonetheless, the mold fiasco later caused problems. After I had applied a second primer coat and a final coat of yellow paint to the hull, the residue of wax caused most of the paint to peel off in large flakes within six months.

At the time, I didn't wonder why a boatyard under intense pressure to make rapid progress on a large-scale production yacht would distract itself with a small sailboat to make some fast money. It wasn't until a month later that this distraction, as well as Chuck's remark about the banker, fell into context as clear as a mountain stream.

⤚

Though we didn't yet have a finished boat, Earl decided when we finished the mold that it was time to start marketing. To this end, he rented space for a table at the Seattle Boat Show **(Illustration 30)** held in the King Dome that January. A sports stadium, the King Dome was probably the largest covered space in the Pacific Northwest. Held annually, the Seattle Boat Show was, and still is, one of the biggest boat shows in the United States. Vendors at the show included nautical publishers; distributors of electronic equipment; manufacturers of survival suits, winches, or nautical hardware; purveyors of commercial fishing gear; outboard and propeller dealers; yacht designers; large- and small-scale boatbuilders; and so on. The King Dome was large enough to accommodate display yachts up to 50 feet long. The sole purpose of the Seattle Boat Show was to sell, sell, sell.

The weather was atypically cold that January, with snow on the ground. We drove daily from Tacoma to Seattle on icy streets for the eight-day duration of the show. Chuck, Earl, Breezie, Griz, and I took turns at the table in pairs, so none of us was there every day. As I rode with Earl to set up our table the first day of the show, I asked him why we were advertising if we didn't yet have anything to sell.

"It's possible someone will like Aquila so much they'll put money down for the first hull, but I doubt that'll happen," he said. "We'll be there for exposure, to distribute our brochures, our logo, and our name. That way, when we finally finish a boat and start racing it, people already will have heard of Aquila.

"Believe me, I know marketing, and this is the right thing to do," Earl continued as we slid to a stop at a red light within sight of the King Dome. He pointed at the behemoth structure. "Look up there. For the last 10 years, I've been selling construction materials, and that was one of my last big contracts. I sold all the roofing for the King Dome. Just imagine how much that brought me in commission!"

The vast floor of the King Dome looked like a cross between the start of a yacht race and a powerboat broker's yard. Sailboats with their sails raised appeared to be gliding across the concrete among new powerboats arranged in rows. The main action of the show was there on the floor, because people came primarily to see boats. Our table was located on a second-floor balcony, overlooking but well away from the center of activity.

We weren't the only table on the balcony, but even among our peers, ours was a pathetic effort. Our main display item was a glossy brochure with Aquila's logo—a stylized cluster of wind gusts—and a description of the vessel's admirable qualities, entirely untested though they were. We also had pictures of the plug, to indicate the size and shape of the vessel and prove we were building it. Our brochure simply could not compete with the demonstration boats out on the floor that potential customers could climb into and imagine themselves sailing.

Having two people at the show freed one of us at a time to wander about. After a couple days, I'd seen everything and sitting at the table was as exciting as a padded cell. As a salesman, I was a failure. I blushed and mumbled whenever someone asked who'd designed the boat, and I didn't know enough about sailing to wax eloquent about the advantages of the design.

To relieve the boredom, I began watching a cute brunette who was tending a table farther down the balcony from ours. Her father built custom wooden yachts in his yard in British Columbia. Wearing a hand-knitted woolen sweater with Scandinavian designs, the brunette had a fetching, wholesome look to her that proselytized the intimate, traditional flavor of the family business. She and her father didn't have a demonstration boat in the show either, but they did have lots of glossy photos of their finished products. They had more to show than we did.

I chatted up the brunette whenever I got a break in the action, which was often. To make conversation, I told her that one of the guys I worked for was a master salesman who had sold the roofing material for the King Dome before launching into a successful boatbuilding career.

"Oh, come on," she said, "how do you know he sold the roofing material for the King Dome? Because he told you that? I mean, I could tell you the same thing, that I'd sold the roofing material for the King Dome, and how would you know otherwise? Sounds like a crock of malarkey to me."

All of this was nitpicking, but she did have a point. Anyway, this verbal kneecapping was less of a setback than when I noticed she was spending a lot of time with a muscular guy from another booth. "He's a boatbuilder," she told me, "and he's a competitive swimmer as well. He asked me out tonight."

The natatory bastard was also from British Columbia, which put me at a severe disadvantage. Near the end of the boat show, I made the mistake of mentioning to Earl my attraction to the brunette from BC.

"Listen," Earl said, "as a salesman, the first thing you have to sell is yourself. For the customer to believe he's getting a first-rate product, he's got to believe a first-rate person is asking him to buy it. If the customer doesn't believe in you, he won't buy what you're selling. I saw you over there with that Canadian chick, and you looked like a puppy dog waiting for a handout. For chrissake, if you don't display confidence in yourself, why should anyone else have confidence in you?

"It's too bad the boat show's over with," he added, "because with a week of coaching, I could have put you deep inside that woman."

Maybe this was just more of Earl's braggadocio. Then again, maybe he really had sold the roofing material for the King Dome. Earl was shadows, mirrors, strings, and fog, and I never knew quite what to believe.

⌒

Early in January 1979 we began laying up the first hull of Aquila inside the mold. Lyall had been laid off, ostensibly because his specialized services were no longer needed but possibly also to cut costs. Chris had also been laid off. The remaining crew comprised Chuck, Earl, Griz, and me. We waxed the inside of the mold and then sprayed it with an off-white gel coat that would become the surface of the hull. Because of the supporting frames, we could tilt the mold while we worked on one side, and then rotate it the other direction to work on the other side. This made fiberglassing easier, because by rotating the mold, the only vertical surface we had to deal with was the transom.

We laid up the hull as solid fiberglass, layer after layer of mat and roving, with around 10 longitudinal stringers running nearly the length of the vessel as stiffeners. Though laying up was straightforward, it was not trivial; the surface area of the hull was at least 700 square feet, or close to 80 square yards. If I remember correctly, it required at least a day to lay down a layer of mat and roving on half the mold, which we left to cure overnight. Before adding the next layer, we roughened the entire surface of the cured layer with a disc grinder to allow proper binding. As with the mold, we worked inside a plastic tent, accompanied by intermittent racket from the furnace. Sucking air through our breathing masks, we sounded like patients in a tuberculosis ward. Sweat poured into our protective clothing and trickled down around our masks. When we left the heated tent to clean our tools, we chilled rapidly from the sweat.

One evening after a long day of frolicking in petrochemical hell, Chuck took me aside. "Matt, I need to talk to you," he said. He looked uneasy and sheepish. "I don't quite know how to say this, but we're going to have to lay you off. The fact is, we simply don't have the money to pay you."

I didn't take this news well. I was dumbstruck and felt mildly angry.

"Jesus Christ, Chuck, you've got to be kidding. You've got tens of thousands of dollars invested in Aquila, and you need to finish a boat to get anything at all back. How can you be out of money?"

"Yeah, I know, but without going into a long, sad story, all I can say is that we've got to cut costs to a bare minimum."

I suppose the noble thing to do would have been to volunteer to work for nothing. After all, Chuck and I had become friends. Over the past year we'd dined together, sailed together, and gone to rock concerts and clubs together. As a friend, I should have offered to help him through a bad time, but this didn't even cross my mind. In truth, it wouldn't have made much difference.

"So, how are you going to finish the boat with fewer people?" I asked.

"Well, Earl will obviously stay on because he's an investor, and Griz will stay on because he goes wherever Earl goes. Griz's girlfriend is going to help us out, too."

Griz's girlfriend, a pretty, doe-eyed brunette in her late teens, seemed not overly ambitious. She'd spent time around the boatyard over the past year, popping Juicy Fruit gum and watching us work while she waited for Griz. Griz had leaked the news not too long before that she was pregnant. To vent some of my frustration, I focused on this.

"OK, let me get this straight. You're firing me and hiring a pregnant woman who knows absolutely nothing about boatbuilding in my place."

"No, it's not like that! Her arrangement will be with Earl. The company won't be paying her, and she can do things that don't require any skill, like cutting material and mixing resin. So, it's not like we're hiring her in your place."

I intentionally ignored this complimentary explanation to get in a final barb. "Well, I never thought you'd be desperate enough to risk exposing an unborn baby to fetal acetone syndrome." I made that last bit up, but it sounded good at the time.

With that, I hefted my toolbox and coverall and trod down Marine View Drive toward the bus stop, feeling intensely sorry for myself. Appropriately, it was cold, dark, and raining. The wheels of the approaching bus made a hissing sound as they threw up water from the road. In those days, the throaty sound of a bus was the loneliest sound I knew—perhaps because most sad partings in my life had involved buses.

I boarded the bus. I reeked of resin and acetone, but since this was the Flats the other passengers didn't notice. After I got home and had a shower, I reflected that I'd miss Knapp Boatbuilding a.k.a. CLK Yacht Crafters. I'd had good times there and learned a lot. My petty anger gave way to sadness for the way things were going for Chuck, and I knew we'd remain friends. Laying me off was probably harder on him than it was on me.

Chapter 12
The sugarcoated *Adelphi* story

A boat is a wood-lined hole in the water into which you pour money.

—Anonymous

Suddenly being laid off a job is always like being doused with cold water, but when you're young it doesn't have the same impact as it does when you have a big mortgage and kids in college. I had enough money for food and the next month's rent. It was actually a relief not to have to trundle down to CLK every day after school and wallow in cold fiberglass resin.

An incident occurred soon after I lost my job, too, that greatly buoyed my spirits. A month or so after I'd moved into my dungeon apartment on Tacoma Avenue the previous September, the landlady had come down from her attic apartment to tell me she was selling the building, and that some prospective buyers would be touring the place. The buyers showed up one Sunday morning. There were three of them, all Italian-looking, all in gray trench coats, all wearing fedoras. It was as though they'd just stepped off the Godfather set, bit players taking a coffee break. They hunched around the basement opening cabinets and inspecting the appliances, all the while talking low among themselves in what sounded like genuine New Jersey accents. They seemed out of place in Tacoma.

"We might be your new landlords," one of them threatened as they departed, and eventually they did become my new landlords. Nothing inside the apartment changed, nor did they raise the rent. But they did evict the first-floor tenant above me and install instead a nephew to oversee the building. I was to pay my rent to the nephew at the start of every month.

The nephew was in his twenties; had short, curly black hair; and wore silk shirts and enough gold jewelry to qualify as a jewelry store.

He fancied himself an Italian stallion, a real Palermo palomino, moving up in the world. He drove a pink Ford Mustang that wasn't a convertible but should have been. Almost every night, he partied somewhere, arrived home after midnight with a great gunning of his car engine, and banged into his apartment with his girlfriend. I knew when he got home because he usually woke me up.

The nephew and I would have gotten along fine if he and his girlfriend had just gone to bed when they arrived home, but they didn't. Instead, they played what I can only conjecture was an ancient Italian bedroom game, the same game several times a week. There'd be a thumping noise, and a pitter-patter, pitter-patter, pitter-patter of bare feet from one end of the apartment above mine to the other. Then silence. Then giggling. Then thump, pitter-patter, pitter-patter, etc. Again and again, this sequence would continue for an hour or more, or so it seemed. The problem was, once it started, I couldn't get back to sleep. Sheer curiosity kept me awake, wondering what they could possibly be doing that would generate this bizarre, repetitive series of sounds.

Finally, after listening to pitter-pattering over a period of several weeks, I knocked on the nephew's door at 1 AM and asked whether he might perhaps try to be a little quieter.

"If youse don't like it," he said, "maybe youse ought to move out. And another t'ing, don't ever bodder me at home like dis again. I deserves a little respect heah."

I thought the nephew could improve his people skills. At any rate, not wanting to move, I endured, and sometimes endurance is rewarded in wondrous ways.

One unusually cold February night not long after I'd lost my job at CLK, I was lying awake listening to pitter-pattering and grinding my teeth, when I heard a hideous wrenching noise out on the street. It sounded like a forklift running full speed into a large stack of sheet metal. I immediately dressed and ran outside in slippers to see what had happened.

What had happened was that a drunk roaring down the avenue had slammed into the nephew's Mustang, converting one side of it to scrap and shooting it like a pool ball into the car parked ahead. The drunk, perhaps not even aware he had hit anything, had continued on down the block. But the collision had done serious damage to his car as well, and it had quit in the middle of the street. I could hear the futile grinding of a starter, and soon several weaving figures climbed from the car and began pushing it—not toward the curb, but in the same direction they'd been travelling.

"Aha," I thought. "It is my civic duty to inform my neighbor that his car has been involved in an attempted hit-and-run."

I climbed the stairs to the nephew's door and pounded on it. I could hear pitter-pattering and giggling inside. It was a testament to the nephew's power of intense concentration that he had not heard his automobile being shredded, and it took a full minute of pounding to get him to the door. The nephew yanked open the door like he hated it, and when he saw me on his doorstep, he flew into a rage.

"The fuck youse want?" he said, "I told youse never to bodder me! I oughta come out 'n' kick your ass!"

"Whoa, whoa, hold on," I said. "I'm really very sorry to disturb you at this time of night, but since you're my neighbor, I thought I should tell you that someone just totaled your car out on the street. I thought you'd want to know."

He looked down at the street. It took him a few seconds to realize it was, in fact, his car that looked like an eviscerated pink pig in the sickly yellow light of a street lamp. He let out a bellow, "Noooooo...," and ran off toward the drunk's car, his bathrobe streaming behind him, his breath like smoke in the cold air.

I hurried back to my apartment, leaving the nephew and the drunk to work things out. It was too late and too bitter outside to stand around gawking. As I drifted off into the sleep of the virtuous, it occurred to me that maybe the drunk knew the nephew and hadn't hit his car by accident.

⌒

Having some free time, I started a boatbuilding project with Perk Haines, who'd been the electrician on *Macs' Effort*. Since he'd left the yard the previous summer, Perk had visited periodically to see how Aquila was coming along and to swap lies with Don. One day in the fall, I'd wandered into the yard of Hylebos Boat Haven and encountered Perk examining every detail of a 24-foot lapstrake dory **(Illustration 31)** that he told me had been built by one of Trumbly's former students, Carl Bronstein. Propped up on blocks, it was a beautiful boat, with an inboard diesel engine and a comfortably large cabin.

When I finally caught Perk's attention, I noticed he had The Look. The Look is what a person gets when he falls in love with a particular boat and knows he must have it, or one just like it. The Look is the result of the person fantasizing about taking that boat to the places of his dreams. After a lifetime of immunity, Perk had finally caught a bad case of boat fever.

That was the origin of our project together. Whenever we encountered one another subsequently, Perk and I talked dories. Although Perk had wired a lot of boats, he'd never built one, and now he was obsessed with the idea. When I lost my job, he was waiting in the wings with a business proposition.

Perk proposed that we build a small dory together so that he could learn the fundamentals of boatbuilding. The deal was that he'd provide a place to work—his garage—and buy all the materials for lofting, patterns, and construction of the first boat. I'd show him the whole process of constructing a hull from a set of lines. When the boat was finished, we'd sell it and share equally in the profit. With a complete set of patterns and a strongback, he could then easily build additional boats if it looked like they would sell, and he could use the proceeds from those eventually to finance his big dory.

Perk was like a Roman patrician who, having survived a lifetime of intrigues in the Senate, had retired to his villa in the hills overlooking Rome. He lived with his wife Lettie in a big, ranch-style house on a bluff in Northeast Tacoma overlooking the Flats. His dining room window provided a superb view of that vast industrial wasteland and the hazy hills of Tacoma beyond. Over his morning coffee, Perk could watch steam rise from the pulp mills and chemical plants, and observe log trucks and chemical haulers crawling like ants in the distance, as in a silent cinema. Whereas Perk was talkative, active, and outgoing, Lettie was a small, arthritic, birdlike woman who spent her days knitting. When I visited, she rarely said more than "hello" and "good-by."

I mentioned to Klarich one day that Perk and I were going to build a dory. Klarich was immediately interested and asked if he could join us. Perk had no objection—the more, the merrier, as far as he was concerned. Though we were entering into a business enterprise, neither Klarich nor I really cared whether we made any money from it. We both liked Perk, and above all we were eager for any boatbuilding experience we could get, especially if someone else was buying the materials.

Rather than design a boat, we decided to build a rowing dory from a book of Phil Bolger's designs. The dory was a good choice, flat-bottomed and simple. However, we discovered when we began lofting that the offsets were in metric rather than English units, which we hadn't noticed when we chose the plan. We thus needed to convert the offsets from centimeters and decimals thereof to feet, inches, eighths, and sixteenths. We couldn't use the offsets as they were, because our tape measures, folding rules, and squares were marked in English units. We made errors in the conversions, but the beauty of lofting was that it exposed these errors straightaway. Nonetheless, to phrase it in the English system, having to convert from metric to English was sodding inconvenient and buggered us on several occasions.

By the time I left Tacoma in late April, we had the hull **(Illustration 32)** constructed, though unfinished; that is, it needed sheer guards and rails, flooring, a breast hook, and thwarts, and it needed painting. Neither Klarich nor I made a dime from the project, and I don't think Perk ever did either.

It is curious that Perk and I got along so well together, for our outlooks on life were very different. Once he told me proudly that he'd just gotten his grandson a job in a chemical factory on the Flats.

"If he keeps his nose clean and works hard," Perk said, "he ought to be set for life."

I nodded my head in assent but was aghast. At best this seemed a sentence to a monotonous and dismal existence, to start in a chemical plant in one's twenties and punch in every workday until retirement. At worst, it seemed a one-way ticket to a cancer ward. Perk, however, had endured the Great Depression, and a steady job wasn't something he took for granted. Admittedly, I didn't know his grandson. Maybe a lifelong job in a chemical factory was the best the kid could hope for.

⤺

One Saturday evening, Klarich stopped by my place and we had a few beers. The weather was soggy and chilly. Klarich said, "Let's go check out the party at UPS," meaning University of Puget Sound in Tacoma.

"What party?" I said.

"Every Saturday there's an open party at the student center: beer, loud music, lots of chicks. You're supposed to be a student there, but no one ever checks IDs. I go up there all the time."

So we gathered up the rest of our beers and drove to UPS in White Cloud, his pickup. Either we got there late or we'd landed on the negative tail of stochastic party variation, because the place was dead. There was only a handful of tired-looking students, and the few women in attendance weren't impressed with a couple of vocational school students noticeably older than they were.

After partaking liberally from a keg of beer, we began meandering around the campus in White Cloud. We were sloshed, but drunk driving wasn't taken as seriously then as it is now. In fact, we'd have had nothing to worry about had not a stop sign jumped out in front of White Cloud and gone soaring over the cab, taking a headlight with it.

"Oh shit, man," said Klarich, "I've got to take evasive action."

This consisted of further meandering around the campus, only slower. Finally, we decided to call it a night and head for home. We were pointed for home when Klarich slowed by the curb and gazed wistfully out the window on my side, like Balboa viewing the Pacific Ocean. I followed his gaze, and it led to a vast, smooth, waterlogged lawn studded with scattered conifers, like a flat giant slalom run. By God, here was the smooth, green belly of Mother Earth herself!

"Shall we?" said Klarich.

I knew instantly what he meant, and it was enticing. I had a drunken vision of White Cloud gliding in and out among the trees, throwing up a big rooster tail of mud and spray, leaving a trail of disdain for the ivy-covered halls occupied by women who didn't appreciate manly workingmen like ourselves. In this vision, Klarich and I were laughing crazily, holding beers out the windows.

So I said, "Sure."

Klarich got White Cloud's front wheels up over the curb and was creeping forward so as not to take out the oil pan, when Conscientious Matt took over from Drunken Matt and fast-forwarded the vision of Men-with-Beer-Having-Fun-in-a-Ford-Pickup to its logical conclusion. The logical conclusion was White Cloud bogged down in the soggy lawn; ugly muddy scars on Mother Earth's belly leading from the curb to White Cloud; and enraged campus cops, alerted by a stop sign in the middle of the street and reports of a ghostly pickup, converging on White Cloud and arresting her occupants. End of logical conclusion: stiff fines, probation, and hundreds of hours of community service doing lawn restoration in lieu of jail sentence.

"Hold it!" I said. "I don't want to do this."

Klarich stopped and stared at me for half a minute. Then he put White Cloud in reverse, backed slowly off the curb, and headed for my place. We weren't more than a few blocks away from UPS when he started in on me.

"You're a pussy, you know that? I thought you were a man, but you're just a pussy, afraid of a few little eensy-weensy campus pigs!"

I explained the Extension of the Vision to him, but he wouldn't let up. Finally, I said, "OK, just stop the fucking truck and let me out. If you want to spend the night in jail, then have at it."

And so he let me out, about a mile from my apartment.

"Pussy!" he shouted, as he sped off, White Cloud gliding like wind down the deserted street.

Our friendship was strained for some time after that. The moment of truth, the opportunity for trial by fire, male bonding, and so forth, had slipped past us unfulfilled. Of course, our growing apart may have had nothing to do with the lawn incident. Soon afterward, Klarich began pursuing a vixen named Betty and wasn't rational for many months. Then again, he was not always rational even in the best of circumstances.

In early February, I'd been out of work for several weeks and was beginning to worry about making ends meet. My little sailboat was reaching an expensive stage of her upbringing; she needed long, clear, vertical-grain Sitka spruce for a hollow mast; she needed sails; and she needed expensive stainless-steel hardware. Patrick Chapman happened to be out of work at the same time, and once again Trumbly came to our rescue.

The previous July, Trumbly'd had a big truckload of Port Orford cedar delivered from Oregon to the courtyard behind the Bates boatbuilding shop. It was a magnificent sight: rough-cut timbers of boat-quality wood, 4 x 4s and 4 x 12s in 30-foot lengths. The class helped unload it and stack it with spacers. I estimated 17,000 to 20,000 board feet, which even then was worth a small fortune. This was Trumbly's personal purchase, which he intended to use for *Windance*.

Trumbly'd had the lumber delivered to Bates rather than his home because he needed to rip most of it into flat strips, 1/4 or 3/8 inch thick and 4 inches wide, to use for cold molding *Windance*'s hull. A cold-molded hull has a higher strength-to-weight ratio than the more traditional carvel-planked hull and requires no caulking of seams. However, there are also disadvantages. Not only is cold molding more labor intensive than carvel planking, but it also involves considerable waste. In ripping strips from solid lumber, nearly 1/8 inch of material is lost with each cut of the table saw. In other words, if one converts 10,000 board feet of lumber into 1/4- or 3/8-inch strip planking, one loses roughly 2500 to 3000 board feet as sawdust. There is also the significant expense of buying epoxy glue by the 55-gallon drum to laminate the strips.

Trumbly was, of course, completely aware of the advantages and disadvantages of cold molding. He chose this technique for *Windance* simply because he'd built carvel-planked boats his whole career and wanted to do something different. No one ever accused Trumbly of being stuck in a rut.

That February, Trumbly hired Patrick and me to work with him after school for a couple weeks. We ripped thousands of board feet of Port Orford cedar into strips on the table saw, then ran the strips through the planer twice to smooth both sides. Finally, we tied the strips into bundles for Trumbly to transport to his home. Not long after we finished that job, he outlined another opportunity for us.

"A guy named Doug White is rebuilding a sailboat in Old Tacoma. Last year, he asked me to take a look at it, and I did. He wanted advice on the best way to proceed. After I saw the boat, I told him the best solution was to buy a book of matches and burn the damn thing. But he didn't take my advice; he said he already had too much money in the project to back out. Now, he's pushing a launching deadline and needs some extra help. I told him I'd send him a couple of my hotshots, and I also told him you guys don't work for nothing.

"Since you two need work right now, why don't you go down there and talk to him? There's no obligation on your part. I don't much like restoration work, and I wouldn't blame you if you refused the job. But, if you can get some experience and a few paychecks out of the deal, it might be worth your while."

This is how Patrick and I came to work for six weeks helping rebuild an old sailboat named *Adelphi*. Our involvement ended with the launching of the vessel in March. Soon after the launching, I submitted an article about the restoration to *Nor'westing*, a monthly magazine focused on sailboat cruising in the Pacific Northwest.

The article was a highly sugarcoated version of events. I had a major journalistic hang-up, which was that I often sacrificed honesty to preserve a sense of propriety. In other words, I was hesitant to hurt anyone's feelings by airing dirty laundry in print, even if it happened to be the truth. Before I submitted an article, I always let the primary source read it. This was putatively to ensure accuracy, but it also imposed a form of self-censorship. I simply didn't have the guts to characterize a guy in print as a fatuous yuppie dink if I was going to let him read the article. This goody-two-shoes approach is a serious defect in a journalist.

As an introduction to the *Adelphi* affair, which phrased in this way sounds like a spy thriller, here is the article exactly as I submitted it to *Nor'westing*:

THE RESTORATION OF THE *ADELPHI*

By the time of her launching (or re-launching, as it were) on March 24, 1979, 70 years after she was built, the 45' sloop *Adelphi* had become an object of wild speculation, gossip and bet making among the Tacoma yachting clique. A year and a half before, she had been dragged from the water onto the Old Tacoma ways for a quick refitting. In a manner common to old boats and corrupt bureaucracies, close examination revealed layer upon layer of rot, from deck to frames to planking, on down to the very backbone. Within six months after she was hauled out, the *Adelphi* stood gutted, the topsides completely removed, and her owners had to decide whether she was worth saving. The most knowledgeable sources suggested gasoline and matches as the practical remedy. Old yachtsmen maintained that she wasn't well built to start with, so why try to save her? Financiers stated that no individual in his right mind could afford to rebuild her. The classic old hull appeared doomed, with public opinion against her.

She would have been, except that a man named Doug White chose her as his. But this is only a late chapter in a story that began in 1907. That year, two brothers in the Royal Vancouver Yacht Club, Claude and Harold Thicke, commissioned naval architect Edson B. Schock to design for them a small auxiliary schooner. Schock later achieved some fame as designer of the *Berneyo*, which won the Havana powerboat race. The *Adelphi* commission was one of Schock's first jobs after he moved to the West Coast from New York. He supervised her construction personally in Vancouver Shipyard, British Columbia. When the cabin was ready for construction, the Thicke brothers and "Eddy" Shock argued over cabin knee placement, resulting in Shock's sudden resignation and furious departure for California. The Thickes had the hull moved to a tugboat yard where the shipwright, in the best of tug-boating tradition, built a straight-sided cabin on the *Adelphi*. The schooner, launched in 1909, was gaff rigged, 31 feet in waterline length, with 11 feet of beam, 6 1/2 feet of draft, and 1100 square feet of sail.

From 1910 to 1915, *Adelphi* was active in Vancouver racing, holding her own in the cruiser class. Apparently she was one of Schock's first commissions and, though pleasing to the eye, never was a record breaker. She placed second in several Beaver Cup races. Usually the famous *Minerva* and the well-known *Uwhila* beat her out of the trophy. There are conflicting reports concerning *Adelphi*'s racing from Victoria to Maui around 1918 and placing in that event. In 1919, the Thicke brothers sold *Adelphi* to Bert Austin, who won the PIYA International Regatta at Cowichan Bay with her in 1921. Austin in turn sold her to James McKee, who retired her from racing in 1922 and moved her to Seattle, where she remained for many years. During the Second World War, McKee lived on the vessel in Eagle Harbor and changed her from a schooner to a cutter-rigged sloop.

After one more change of hands, *Adelphi* was bought by Charles Udall, a well-known Tacoma contractor and developer, in 1957. Udall and his family raced her extensively in Tacoma Yacht Club races during the late 1950s and early 1960s. For most of the 1970s, the aging vessel sat idle in her slip, and in 1977 Udall decided to sell her. He tacked a notice to that effect and a photograph of the boat to the yacht club bulletin board, an event that changed the life of Doug White and lifted a sentence of death-by-slow-rot from *Adelphi*.

When Doug White bought *Adelphi*, he had just left a successful career in civil service to enter business on his own. For several years he had been remodeling antique houses in Tacoma, more or less as a hobby. Once he made the decision to jump off the bureaucratic mill, he founded the Property Development Corporation, dedicated to the restoration and preservation of old houses. In the past year and a half, White has completely or partially restored 11 houses in Tacoma, and he proudly boasts of living in that city's oldest (1883) Victorian, which he and his family restored from a condemned hulk to a museum piece. All of this has a bearing on the *Adelphi* story. White had always wanted a boat, and when, during a period of self-fulfilling personal renaissance he decided to buy one, it was natural he choose a restoration project.

Had White known the magnitude of the project he and his first partner, Dick Hayes, were becoming involved in, he would probably not have undertaken it. As he stripped off the deck canvas and old paint, he found rot everywhere: in the deck and cabin beams and planking, sheer clamp, upper planking, frames, floor timbers, horn timber, and so forth. The keel bolts were nearly rusted through. White realized he would have either to write *Adelphi* off as a loss or rebuild her completely. His partner chose to bail out, which left White holding the most expensive and controversial wreck in Tacoma.

Accustomed to the rigors of restoration, White opted to continue. He hired local shipwrights and jumped into the work with incredible energy. The entire deck, cabin, and interior, and some planking and frames were stripped away, so that early in 1978 not even a complete hull remained. At this stage, a very lucky thing happened. White met Steve Thrasher, a mild-mannered professor in the School of Business and Finance at the University of Puget Sound. Thrasher took an interest in *Adelphi* and become a partner in the restoration **(Illustration 33)**.

And so the work progressed. Shipwrights Ron Langer, Mike Vlahovich, and Phil Tate, all ex-students of the boatbuilding program at the Bates Vocational Institute, and old-timer Clarence Potts, worked on the structural carpentry during the daytime (though not all simultaneously). White, his son Steve, and Thrasher came in on the evening shift to sand, caulk, paint, plug, and finish.

The *Adelphi* that slid down the ways this March was a somewhat different, and perhaps better, boat than the one launched in 1909. Two thirds of her planking is new, as are her horn timber and everything from the deck up. The original floor timbers were 3-inch-thick fir every 20 inches; now they are 2-inch-thick oak every 9 inches. The original stringers were of weak 2-inch by 3-inch red cedar; now they are of strong 2.5-inch by 4-inch yellow cedar. The original canvassed decks have been replaced by laid Honduras mahogany with caulked and rubbered seams. Whereas the original chain plates were bolted directly to the planking, the new ones are bolted to a stringer running along several frames, and reinforced with aluminum deck knees. The straight cabin sides have been replaced with beautiful, curved, solid, Honduras mahogany ones. The vast cockpit has been shortened to safer and more modern proportions. The new bowsprit is half the length of the original, giving the center of effort approximately 17% lead. In spite of the changes, *Adelphi* retains her old, graceful look.

There is no doubt that the efforts of White, Thrasher, and their families saved the *Adelphi*. Conversely, the restoration has had a profound effect upon the participants. For them, there was something more important than collapsing in front of a television at the end of a day. *Adelphi* would not wait: shop rental fees accumulated; workers demanded materials; the day of the Daffodil Regatta, which *Adelphi* was to lead, drew near; and month by month, materials became more expensive.

Adelphi came along when Doug White was undergoing a radical personal change, and the two grew together. "When you sit at a desk and shuffle papers," he is fond of saying, "you have nothing concrete to show for it. When your kids ask, 'What did you do today, Dad,' you have difficulty explaining to them. However, if you built a hatch cover or laid a deck, you can point and answer: 'Why, that's what I did today.'"

Doug's wife Judy, who delivered many a near-midnight dinner to her husband and his co-workers at the Old Tacoma shop, is enrolled in sailing lessons in anticipation of the summer's cruising. Fourteen-year-old Stephen White cut his teeth on quite an assortment of tools and techniques during the project. In the Thrasher family, Jennie was the experienced sailor, having grown up on the Great Lakes. She did not begrudge her husband the hundreds of hours he labored on what seemed to be an interminable project. And Steve Thrasher deserves the medal for patience and valor. Day after day he stuck to thankless tasks like rubbering the deck, plugging, and sanding. He smoothed out tiffs between his volatile partner White and the workers, and infused a spirit of good-natured cooperation into the project. The carrot at the end of his stick was the day when he would sail aboard *Adelphi*.

The *Adelphi* launching was a festive occasion with beer and cheering, a rite of spring. Most of those present knew of the sacrifices made by the two families and could join in their exuberance as the boat slid down the ways and headed for her slip in the Tacoma Yacht Club under her own power, like an old mare to the barn. Someone asked Doug White, "What's next?" Why, of course, he's thinking of taking on contracts for the restoration of historic boats.

The End

By the time I received a reply from the editor, I was back in Kodiak. The editor was no fool; he went right to the heart of my journalistic hangup. The article was okay, though tedious, in presenting the history of *Adelphi*, but it obviously degenerated into sentimental schlock at the end, when everyone lived happily ever after. It had the flavor of a 1960s sitcom, where Mom, Dad, and the kids live in blissful denial of 10,000

Soviet nuclear warheads aimed in their general direction. Here is the editor's letter to me:

Nor'westing
July 6, 1979

Dear Mr. Dick,

Forgive me for taking so long to get back to you on the *Adelphi* article. I just found it buried under a pile of papers. I like the idea, but the presentation was a little unconvincing. Could you rewrite the story, maybe putting it in the first person? Tell it from your involvement. This gives you a chance to tell what the people were like, how the boat looked to you, what it was like working on it. It's hard to write a 3rd person report such as this one and make it come off. Much easier to get yourself into it.

Also, the more specific information you provide, the stronger the story is. You were best when you recounted one fact after another, and you were less convincing when you generalized, such as when you talked about a change of lifestyle, or bringing hundreds of meals to the boatyard. If you could find a book by John McPhee (he's written 13, all of them excellent), you'll see what I mean about using specific detail to tell the story. He wrote Coming into the Country, which is about Alaska. I'll bet it can be found in Kodiak. You can't possibly imitate McPhee, but he shows a way.

Hoping that you will take on a rewrite, I'm holding the pictures and the copy of the original article about the boat. I'd like to run something about the project. If you want the pictures back, let me know, and I'll send them, and give up the idea of running something. I realize I've thrown quite a lot at you here, and we've not had correspondence before. I hope that's all right. Please let me know if you can do a rewrite.

Sincerely,
Robert Hale, Editor

I was indignant at the letter. I'd never before had a request for an extensive rewrite, and I bristled at the editor's suggestions. If he wanted the article written a certain way, he could damned well write it himself! The suggestion that I should follow the path of John McPhee irritated me almost as much as the blanket assertion that I couldn't possibly imitate the man. As far as I was concerned, McPhee was a pretentious New York hack who presumed greatly in fostering the illusion he'd captured the essence of Alaska after spending a few weeks there.

And Jesus, did this editorial swine think we lived in igloos in Kodiak? After all, the town did have a bookstore and a library, and there were even a few people there who knew how to read. Was there any doubt *Coming into the Country* could be found in Kodiak? Who did this pissant think he was dealing with, anyway? For all he knew, I might be the next William bloody Shakespeare, an uncut diamond emerging from the wilds of Alaska, who would one day crush no-talents like him. Needless to say, I didn't rewrite the article, and it was never published.

In my heart of hearts, though, I knew Mr. Robert Hale was right. That he'd gone out of his way to try to help me was salt in the wound. The only element of his letter I could honestly quibble with, other than a condescending tone, was his choice of a model. Considering the tenor of my involvement with *Adelphi*, I'd have chosen Hunter S. Thompson's *Fear and Loathing in Las Vegas* after which to style my article. Unfortunately, I couldn't imitate Thompson either.

Chapter 13
The real *Adelphi* story

A job needing considerable care, though basically very simple, is boring through the hull for the shaft.

—Michael Vernay, *Complete Amateur Boatbuilding*

What follows, Mr. Robert Hale of *Nor'westing*, is the story of the restoration of *Adelphi,* narrated in first person from the point-of-view of my involvement. It's as close to the truth as I can get now, through the filter of time. I doubt you'd have published this version, either.

With a full head of brown, shaggy hair; a short, scraggly, salt-and-pepper beard; and big-lensed glasses with a dark plastic frame, Doug White had an owlish look to him. Somewhere in his forties, he typically wore a V-necked sweater, bell-bottomed pants, and loafers. He was a little on the soft side, not what one would describe as fat, but borderline pudgy from living the good life. He said he'd been city manager of Tacoma until recently, when he'd resigned to pursue his own business interests. I had no reason to doubt his stated motives for leaving his job, though one always suspects the machinations of local politics in the resignation of any city manager.

Though Doug claimed to have restored a number of Victorian houses, the bulk of his work when I knew him seemed to comprise smaller remodeling projects, particularly bathrooms. I asked how he could maintain what appeared to be a very comfortable lifestyle doing bathroom carpentry.

"It's not just bathroom carpentry, or bathroom remodeling," he said, "it's status. Look, when I was city manager, I got invited to high-society events and interacted with a lot of shakers and movers in Tacoma. For a while, I walked their walk, and I still talk their talk. After I left city service, an acquaintance knew I was remodeling houses and asked me to do his bathroom. He liked the job I did and spread the word. Now I'm in vogue; it's very 'in,' very chi-chi, to have Doug White do your bathroom.

"Any competent carpenter could do the jobs I do, and probably better than I do them. But I'm not just any competent carpenter. I'm the former city manager of Tacoma. And this is where status comes in. If people hire a person who has a certain amount of status to do a menial job for them, this gives them even more status. The more they have to pay for it, the better.

"A job I'm doing right now is a perfect example of what I'm talking about. A guy is paying me five thousand dollars plus expenses to remodel his bathroom. I act like an eccentric artist, give him just the right amount of high-class verbal abuse in language he can understand. If he asks for green walls, I argue for pink walls, tell him he has no sensitivity. I'm a real pill. The guy has more money than he can spend and literally nothing to do with his time. Haggling with me over his bathroom is a bright spot in his otherwise incredibly boring day. The guy's happy. He loves me.

"Plus, in terms of making a living, it doesn't hurt that my wife sells real estate," he added, almost as an afterthought.

Doug's self-assessment sounded farfetched, but I got some independent confirmation. A student named Joe Brezlin, who'd started the Bates boatbuilding program a few months previously, had a fair amount of experience in house construction. Patrick and I introduced him to Doug, who hired him part-time on remodeling jobs. We considered it beneath our dignity to work on mere houses.

"Doug White doesn't know his ass from a hole in the ground when it comes to interior finishing," Joe told me. "I mean, he doesn't have the slightest idea about current techniques, and a lot of his work is substandard. His clients are satisfied, though, because they don't know their asses from a hole in the ground either."

For our first evening of work on *Adelphi*, Patrick and I followed Doug's directions to a classic old boatshop located on Ruston Way between Old Tacoma Pier and the abandoned Dickman Lumber Company, not far from McCarver Street. This had been my neighborhood the previous year, and I'd passed the shop countless times without really noticing it. The part of Tacoma around the base of McCarver Street is now known as "Old Town." At the time of *Adelphi*'s launching in British Columbia in 1909, Old Town was an active industrial waterfront choked with wharves, sawmills, and boatyards. By 1979, the industries had closed or moved to the Flats, the neighborhood had degenerated into a bedroom community, and Ruston Way was virtually deserted. The shop itself, built in 1938 as a boat repair and sales facility named Tacoma Boat Market, had sat vacant—which was why Doug could afford to rent the place to rebuild *Adelphi*.

Doug and his workers had pulled *Adelphi* up the ways and blocked her up in the cavernous shop **(Illustration 33)**. It was a fitting place for her, because both she and the shop were in about the same state of disrepair. Now, *Adelphi* was faring better. The aging hag had gotten more than a mere facelift; she'd had major reconstructive surgery and was on her way to recovery. Most of her frames were new. She'd been completely re-planked from the waterline down, and scattered planks above the waterline had been excised like the cancerous rot they were and replaced. Freshly caulked and painted, her white hull gleamed like a silk wedding dress. Her new mahogany deck was as rich as a Bermuda tan, and the redesigned, nearly complete cabin top sat on her like a new hat.

The first job Doug assigned Patrick and me was to install low bulwarks—vertical guards 6 inches high running the length of the vessel on each side at the edge of the deck **(Illustration 34)**. To do so, we would make patterns for the curvature of the sheer and cut out the bulwarks in sections from 2-inch-thick lumber, beveling the bottom edge of each to fit the camber of the deck. We would cut out scuppers to drain the deck. We would bed the bulwarks in Sealant 5200 and through-bolt them to the sheer clamps inside the hull, scarfing the sections together as we went to give continuous pieces running fore and aft. Finally, we would shape ironwood caps and attach those atop the bulwarks.

Extending from the boatyard was a covered wooden pier. The power machines Patrick and I needed for the bulwarks were located in a small shop on the pier, separated from the boatyard by a locked gate. Our first day on the job, Doug showed us where the machines were and where to find the key to the gate.

Several journeyman boatbuilders, including three of Trumbly's former students, had done the bulk of the work on *Adelphi*. In fact, two of them were still working days on the vessel. Since we didn't arrive before 4 PM, and they knocked off at 5 PM, we overlapped with them for at most an hour. Often we came in later in the evening and didn't see them at all. This was fine with us, because it meant fewer people in our way as we worked.

By our second or third day on the job, we'd planned out the bulwarks and carried the necessary lumber onto the pier to begin shaping them.

We'd no sooner started up the big band saw when Mike Vlahovich, one of the Bates graduates, came out and signaled he wanted to talk.

"Listen," he said, "I know you guys are Trumbly's students, and believe me, I'm glad you're working and getting some experience. I went through the Bates program myself, and I always like to support it. But, the fact is, you're using equipment out here that you have neither the permission nor the right to use. This equipment belongs to Clarence Potts, and Ron and I have a deal going with him to use it." Ron was another Bates graduate.

"What you've stepped into here is a very touchy issue," Mike continued. "Though Doug no doubt told you he hired you because he's got a launch deadline, it's no coincidence that he can pay you as students a lot less than he pays us as journeymen. What it amounts to is that you're undercutting us. Now I don't blame you, because you wouldn't have known that when you took the job. But Doug's also trying to get some use of this equipment for nothing, and that really pisses me off. He's a cheap bastard; give him an inch and he'll take a mile. You'll see, once you get to know him!"

Patrick and I listened with our mouths agape. There wasn't much we could say.

"Use the equipment tonight," Mike said, "but tell Doug he's got to make an arrangement with us if you want to keep using it."

We all said sorry, no hard feelings, thanks, and goodnight, and Patrick and I continued our work on the bulwarks. When Doug showed up later in the evening, we relayed Mike's message to him.

"Those guys are a bunch of prima donnas," he said. "With them, it's all take and no give. Bloodsuckers is what they are, and they're bleeding me dry. Well, they hold all the cards right now; the bottom line is that I've got to get this boat done. I'll talk to them and straighten things out."

This he must have done, because we continued to use the equipment on the pier and never heard another word about it. The whole incident was a classic labor-management dispute in miniature, apparently peacefully resolved through negotiation.

⤷

Anytime you screw anything wooden to a boat, you countersink the screw. You use a tapered bit that drills a hole the length and diameter of the screw, exclusive of the threads. If you don't predrill screw holes in hardwood, the screws will either split the wood or require so much force to fasten that the heads will strip or the shafts break. The tapered bit has an attachment that simultaneously bores a cylindrical countersink hole a little larger in diameter than the screw head. The head of a fastened screw lies in the countersink hole, below the surface of the wood.

To hide and protect the screw head, you glue a cylindrical wooden plug into the countersink hole **(Illustration 35)**. The plug fits very snugly and usually requires tapping in with a hammer. You take care to orient the grain of the plug in the same direction as the grain of the wood being plugged, so that the trimmed plug will be almost invisible.

Plugs, which have the grain of the wood running transversely, are different from pieces of dowel, which have the grain running longitudinally. Some people who see plugs in the fine furniture they buy think the furniture is put together with pegs, or dowels, but they're looking at plugs covering cheap steel screws.

You make plugs from scraps of the same wood you're plugging, using a type of bit called a plug cutter in a drill press. The plug cutter removes a circle of wood from around each plug. After you've cut many plugs in a single board, you free them all by turning the board on edge and sawing through their bases with a band saw.

After you've set a plug, it protrudes from the wood it's embedded in. When the glue has dried, you must trim the plug flush. This requires a chisel with a slightly curved cutting edge that is literally razor sharp. You first slice off the top of the plug to determine which way the grain is angled. The grain of a plug is rarely absolutely parallel to the surface of the wood. If you attempt to trim it from the wrong direction, it will split off below the surface, leaving an unsightly scar that defeats the whole purpose of the plug.

Once you've broken off a plug, you've got a time-consuming problem. You can't drill out the plug because the screw head underneath will damage your bit. If the wooden surface you're plugging is to be painted, you can use a putty of sawdust and glue to fill the scar, holding the mixture in place with masking tape until the glue sets, and then sand the surface flush. This is not practical if the wood you're plugging is to have a clear finish, because the glue and sawdust mix will show as a blemish lighter than the wood itself. Sometimes you can carefully dig out the damaged plug with a pocketknife and put in another plug, but this is difficult to do without enlarging the countersink hole.

Doug, his teenaged son Steve, and his partner Steve Thrasher frequently worked evenings at the yard, as I mentioned in my article, performing simple tasks like plugging, sanding, and painting. Young

Steve, attracted to the element of danger in wielding a razor-sharp chisel, preferred trimming plugs to anything else. At first Doug wouldn't let him, fearful he'd slash a wrist or put out an eye. This is a common attitude of parents, protecting their children from dangers by not allowing them to learn the skills needed to evade those dangers. The thing is, anyone who uses a chisel eventually cuts himself. This makes him more careful in the future.

The task of trimming plugs initially fell to Steve Thrasher. He had a difficult time learning to read the grain of the wood and tended to lob off plugs in an unfortunate manner. Even after he understood grain, he had trouble sharpening and controlling the chisel. I believe using tools requires the same sort of neural circuitry as playing the piano, and the younger you start, the better you will become. Eventually, realizing that plug trimming was going to take a long time the way it was going, Doug relented and let his son participate. Being only fourteen, young Steve picked it up in a heartbeat.

Plugging is a useful but truly minor aspect of carpentry; it is a mere footnote. I describe it in detail as an example of the multitude of techniques we learned at Bates simply by doing, without realizing we were learning. Back then, Patrick and I could trim hundreds of plugs in short order, without ruining one. To learn this skill, Trumbly or some other student had showed us how to do it, just once. We then went to work plugging the decking and planking on the Trumbly-38, simply getting the job done, making mistakes. The first time I broke off a plug, Trumbly explained what I'd done wrong and the possible ways to correct the error.

"Now fix the sonofabitch," he commanded with a sympathetic grin as he rushed off to another crisis.

Though I didn't realize it at the time, mistakes are not failures. They are instead the very currency of learning. One can read about plugging, one can study it, but nothing focuses one's attention on wood grain and the consequences of a dull chisel better than breaking off a plug. Trumbly was the superb teacher he was because he understood the fundamental role of mistakes in the learning process.

⤶

As I wrote in my article, Doug's and Steve's wives did sometimes bring their husbands dinner at the boatyard. As *Adelphi*'s launching approached, however, there was mounting tension underlying these conjugal sacrifices. Mrs. Thrasher seemed increasingly irritated with the whole project in general and with Doug White in particular. And why wouldn't she be? Her husband was absent from home most evenings and weekends. He was gaunt and distracted from holding down what amounted to a second job in addition to his day job as a professor. What Doug had probably presented to the Thrashers as a fun-filled investment opportunity was turning into domestic and financial hell. Certainly there is no better way to hemorrhage money than to undertake a boat restoration. Journeymen boatbuilders don't come cheap, nor do fine woods and bronze fastenings.

One evening, Steve and his wife engaged in a vituperative whispered spat in a corner of the shop. She stormed out, and Steve followed soon after. Neither of them reappeared for a week. Rumors flew rampant among us workers. Steve's wife was filing for divorce. *Adelphi* had bled the Thrashers dry, and they were filing for bankruptcy. Doug was suing Steve. Steve was suing Doug. Doug was humping Steve's wife. Steve was humping Doug's wife. Doug and Steve were drug-addicted homosexual lovers. All of the above. None of the above. When we ran out of rumors, we speculated upon further possibilities, thereby generating more rumors.

Whatever the nature of the tiff, Steve returned and saw *Adelphi* through to launching. Maybe it was no more than an argument over whose turn it was to do the dishes, but I don't think so. The fact is, Doug White could be difficult to deal with. He wasn't inherently an unkind man, and he could be personable and funny. By the very nature of the project he'd undertaken, however, his financial ass was bared to the wind. Doug's coping behavior was to bluster his way around or over obstacles, including people. At times he seemed a petulant, ham-handed boy trapped in a middle-aged body, oblivious to the negative reactions he generated in people around him.

Near the end of the project, when I interviewed Doug for the article, I asked him if there existed any old photos of *Adelphi* under sail. "Sure," he said, "let's go for a drive." We dropped in on Charles Udall, *Adelphi*'s previous owner. Mr. Udall was not happy when we appeared on his doorstep.

"Jesus, Doug," he said, "don't you think you could pick up a telephone once in a while and maybe call before you come?"

Udall lived in an immaculately maintained, white-stuccoed house perched on a wooded ridge in North Tacoma. The dwelling looked like it belonged in Los Angeles; it had all sorts of fancy little pointed roofs, as if a giant had wrapped his hand around it and squished it messily upward. The tiny yard was meticulously landscaped.

166

Inside, the décor narrowed down my impression of the place specifically to Beverly Hills. Sun streamed into the living room through French doors with a view of Commencement Bay off in the distance. There was not a mote of dust anywhere. Carefully placed Persian rugs protected a highly polished hardwood floor, with thigh-high Chinese vases scattered around the periphery. Antique hardwood wainscotting decorated the walls. None of the furniture looked like it came from a discount store. Udall either had a lot of money or had gone to a great deal of effort to foster that impression. I suspected the former.

A tall, lean, distinguished-looking, white-haired gentleman, Udall held himself straight like a soldier. Retired from a highly successful business, Udall Construction, he still radiated a great force of will. One thing was sure. Udall wasn't the sort you person you just dropped in on. And this is what I mean about Doug. Either he was oblivious to Udall's displeasure or just didn't care. He behaved as though he'd dropped by a tavern on a Saturday afternoon.

"Charlie," he said, "Matt here is writing an article about the renovation of *Adelphi*, and we'd like to borrow some of your photos showing her sailing in the old days."

Udall didn't seem the sort of person you called "Charlie," either.

"Since you bring up the topic of photos," Udall said, "what about those of mine you still have, the ones you borrowed for construction details? You said you'd need them for a week or two, but you've kept them for a year. Are you really going to return them, or are you planning on stealing them? I've been thinking of calling the sheriff."

"Of course I'm not stealing them," Doug said. "I'll get them back to you. It's just that every minute has been occupied with the project. Also, I've had to keep referring to the pictures as the work has progressed."

"Well," Udall said, "if you needed them for a year, you should have borrowed them for a year. If you needed them longer than you borrowed them for, you should have had them copied. One thing is sure; you're not getting any more until I get those back. In fact, I don't think I'll lend you any more pictures at all."

"Okay, look, I apologize," Doug said. "I'll bring them by tomorrow, I give you my word."

"I already had your word I'd get them right back, so I'm not holding my breath. But, if you do happen to bring them by, just slide them through the mail slot."

Thus it was that I didn't get access to any historic action photos of *Adelphi*, though it didn't matter, since my article wasn't published. Doug did manage enough groveling that Udall showed me some photographs in an old album. There were shots of *Adelphi* in her berth, featuring a pretty woman with long, brown hair and a much younger Udall drinking cocktails; occasionally other couples were present. Clearly, *Adelphi* and the Tacoma Yacht Club had once been the focus of Udall's social life.

As we left, Doug reminded Udall that *Adelphi*'s launching was scheduled a few weeks hence.

"I'll check my calendar," Udall said. "If there's nothing important going on, I might be there."

⌒

Our first day on the job, Doug had mentioned that one thing he wanted Patrick and me to do was drill a hole for *Adelphi*'s rudder shaft. This sounds easy, but in fact it is quite challenging. If you've ever used an electric drill on wood, you'll understand why. Drill bits tend to wander. Unless the wood is very thin, the bit rarely emerges on the other side exactly where you intended. The greater the thickness of wood relative to the diameter of the bit, the greater the likelihood the bit will wander. One cause of bit wandering is that it's difficult to accurately judge the position in which you must hold the drill to make the hole emerge where you want it. Furthermore, a bit can bend during drilling. Harder portions of the wood will deflect it toward softer portions, and before you know it, you're slightly askew in direction.

Shaft holes on most boats go right through the backbone, and they must do so accurately. Consider, for example, a propeller shaft hole. A vessel with only a single engine has the engine positioned exactly on the midline. The long steel shaft connecting the engine to the propeller must also run exactly along the midline. If it angles even half an inch to one side at the propeller end, the vessel under power will tend to be propelled slightly toward the same side. This in turn will require constant tension on the wheel to correct the tendency. It's the same thing as driving a badly aligned car that drifts to one side or the other.

Drilling a propeller shaft hole requires that you start the hole exactly on the midline inside the backbone, bore straight through one to several feet of solid hardwood, and exit exactly on the midline outside the backbone. If a long shaft hole emerges an inch or two off center, you've got a big mistake running through the very backbone of your vessel.

It is entertaining to consider how to repair a botched shaft hole so you can re-drill it. With a lathe you turn a long dowel of slightly less diameter than the hole and of the same type of wood as the backbone. You coat the dowel and the hole liberally with glue, and then attempt to pound the former into the latter. But the hole turns out to be slightly curved due to bending of the bit, and the dowel binds tight halfway through. Now, like a dog stuck in copulation, you can't withdraw it either.

This isn't so bad, you think; all you need to do is fill the other half of the hole. You saw off the dowel flush with the backbone and try to pound it in from the other end. But while you've been picking your nose trying to figure out what to do next, the glue in the hole has begun to set. The thickening glue and the air pocket in the middle prevent the remaining half of the dowel from going all the way in. You end up with the hole plugged at both ends, but with a cavity in the center of the backbone.

Thoughts of corrections like these are so unpleasant that boatbuilders take great care in drilling shaft holes. Books on boatbuilding describe various kinds of jury-rigged wooden jigs used to guide the long auger used to bore a hole. Frustrated with jury-rigs, however, some unsung hero in the Tacoma boatbuilding industry vastly improved on the general idea by inventing a set of steel clamps with circular shaft bearings attached to them **(Illustrations 36, 37; also see Illustration 20)**. This adjustable jig eliminated much of the uncertainty involved in drilling shaft holes. Once an approximate guide hole has been drilled through the backbone, the jig can be adjusted to position a cutting rod exactly along the desired centerline of the shaft hole. The beauty of the method is that the guide hole itself need not be absolutely accurate.

Inexplicably, no one manufactured these shaft-boring jigs. There was no place to purchase them. The concept had been transmitted by word of mouth or rough drawings, and the few boatbuilders who owned them had them custom made at machine shops. The jigs were thus exceedingly rare and closely guarded by their owners.

Relevant to the *Adelphi* story is that Trumbly had a shaft-boring jig for use at Bates. Furthermore, he allowed current students to check it out for a day or two at a time for use outside school. This policy did not extend to ex-students, who were too numerous to accommodate and over whom Trumbly had no leverage if they did not return the jig.

Since some of *Adelphi*'s rotten backbone timbers had been replaced, she needed a new rudder shaft hole, the drilling of which involved the same difficulties as a propeller shaft hole. When Doug White had mentioned offhandedly early on that he wanted Patrick and me to drill a shaft hole and mount the rudder, we'd thought nothing of it. Surely he would assign the job to the journeymen instead. We hadn't considered that we alone had access to the Bates shaft-boring jig, whereas our journeyman colleagues didn't.

Somehow Doug knew of the Bates jig and Trumbly's policy governing its use. A couple of weeks before the launching, he mentioned to Patrick and me that it was time we work on the rudder. When we finally realized Doug was serious, we panicked. Neither of us had done a shaft hole before. True, we'd observed the drilling of the rudder shaft hole for the Trumbly-38 and the propeller shaft hole for *Windance*. These operations had gone flawlessly, but of course they would have, because Trumbly oversaw the former and did the latter himself. Thoughts of everything that could go wrong scared us spitless, but we had an idea. We'd borrow the shaft-boring jig, and the journeyman former Bates students working on *Adelphi* could do the job.

"No way," Mike Vlahovich said when we presented our solution to him. "I knew from the start that one of the reasons Doug hired you was that you had access to Trumbly's jig. I told you he's a devious sonofabitch. Well, I'll be damned if we're going to play into his hands. You took the job claiming you could bore a shaft hole, and now by God you've got to do it. You're on your own."

So Patrick and I took the only rational course, which was to stall, hoping Doug would find someone else for the task. On the Friday a little over a week before the launch date, we arrived at work directly after school. Doug was waiting for us.

"When are you going to mount the rudder?" he said. We hemmed and hawed, mentioned something about having forgotten to check out the jig, and said we'd do it next week.

"You guys don't understand," Doug said, starting to heat up. "We're out of time here! Launching is a week from Sunday, and that rudder has got to be mounted now! What if you screw up? I can't take the chance that a week from now, you'll still be fiddling with the thing."

Time and tide wait for no man. They don't wait for women either, or boats. The dates and times for boat launchings are determined by so-called spring tides, which occur twice every lunar month the year around. On exceptionally high spring tides, the sea near shore will be deep enough to float a boat from the ways. If you miss a launch date, you might have to wait another month to get the boat into the water. Patrick and I had both lived in Kodiak and were well aware of the inflexible nature of tides and launch dates.

"Geez, Doug, we're sorry," Patrick said, "but Bates is closed until Monday. We won't be able to get the jig until then."

"No," Doug shouted, frothing at the corners of the mouth. "I can't accept that. You promised you'd get the jig this weekend. So go get it. I don't care what it takes. Find a janitor to let you in, blow him if you have to, break in, I don't give a shit. I want that rudder on this weekend, period! You promised, goddammit!"

Neither of us remembered making any promises at all, let alone concerning this weekend in particular, and Doug's belligerent attitude provided an excellent excuse for us simply to quit. But, the fact was, we hadn't forgotten the jig. We hadn't checked it out because we were afraid of boring the shaft hole. This alone shamed us into returning to Bates.

By incredibly good luck, Trumbly was still there. Usually he rushed home after school like a cow on fire, but this Friday he'd stayed late to rip some fairing battens for *Windance*. We were able to get the jig and the big drill motor used to turn the cutting rod. Trumbly spent half an hour reviewing for us the process of boring a shaft hole.

"The only thing you have to fear is fear itself," Trumbly remarked as we left. "Remember, there's no mistake you can make that can't be corrected." That wasn't the most encouraging thing he could have said right then.

↫

In 1891, an American genius named Nathanial Herreshoff designed and built a yacht he christened *Gloriana*. With a long, sweeping stem lacking any trace of a forefoot, an extended overhanging stern, and a long, narrow keel with external lead ballast, *Gloriana* was so odd-looking compared to the norm of the times that Herreshoff's friends and foes alike predicted she would be a disaster. They were wrong. *Gloriana*'s stability, speed, and ability to sail close to the wind astounded the yachting world and revolutionized yacht design. *Gloriana* became the prototype for the modern ocean racer.

Herreshoff's competitors immediately began to imitate *Gloriana*'s features but failed to understand the principles behind her. Herreshoff was so far ahead that he designed and constructed every America's Cup winner over a span of 27 years.

Like most *Gloriana* wannabes in the early 20th century, *Adelphi* had a long, deep, narrow keel with external lead ballast. The tall rudder spanned the entire aft end of the keel but had a deep semicircular notch cut out toward the top of the leading edge to allow space for the propeller. This type of rudder is called a "keyhole rudder" **(Illustration 38)**.

The hole for the rudder shaft needed to emerge through the horn timber of the backbone exactly on the midline, slightly behind and exactly parallel to the aft edge of the keel. Since *Adelphi* was completely planked, drilling from the inside would have been like trying to hit a bullseye blindfolded. Fortunately, the straight aft edge of the keel provided a perfect line of reference, along which Patrick and I set up a makeshift jig as a means of drilling a guide hole upward through the horn timber.

Doug treated us as experts who knew what we were doing, which meant he left us entirely alone. This was good. Our Bates brethren luckily were not working on the Saturday we started. Patrick and I lined up the guide bit to the best of our abilities and then jumped off into the abyss as we drove it up through the horn timber. I remember a black feeling as the guide bit emerged enough off center on the inside of the horn timber that it was a cause for serious concern. We pretended everything was going flawlessly, but we immediately called Chuck Knapp for advice on how to salvage the job.

Chuck was busier than a four-armed pickpocket, but like a crossword-puzzle addict he simply couldn't resist a good boatbuilding challenge, which this clearly was. He not only advised us but also spent most of the afternoon actively participating. Chuck had both an intuitive sense for problem solving and a way of defusing stressful situations with humor. For example, if you swore roundly at some mishap, Chuck might gaze at you seriously and say, "I'm sorry, but I'd appreciate it if you'd clean up your language."

If you didn't know him very well and apologized, he'd add, "Always remember: profanity is the crutch of the inarticulate motherfucker." You had to laugh.

With Chuck's help, we finally bored the shaft hole straight and true. This, however, proved to be almost insignificant compared to mounting the rudder, which was a massive wooden crescent two inches thick; it was taller than I and weighed 150 pounds if it weighed an ounce. Fitting it was like wrestling a bear, and the task took us 10 hours of feverish work on Sunday.

First we installed the rudder port, a length of pipe running from the deck through the stern overhang and bedding firmly into the horn timber. The function of the rudder port was to provide a watertight housing around the rudder shaft. We then mounted the deck bearing that would hold in place the upper end of the rudder shaft. We inserted the rudder shaft from the keel side upward through the rudder port and into the deck bearing. At the foot of the keel aft, we bolted a stout, stainless-steel heel plate with a gudgeon socket, into which we dropped the end of the rudder shaft.

With these components in place, we propped up the rudder and bolted it into two U-shaped steel straps welded to the rudder shaft. We then inserted two U-shaped hinge straps around the hinge rod running along the forward edge of the rudder and bolted them to the keel. Finally, Sunday evening, we attached the tiller to the top of the rudder shaft that extended above the deck bearing. When we tested the rudder, it swiveled like greased silk.

Despite the imminence of the launch date, Doug wasn't in the shop the weekend we worked on the rudder. It was as though he couldn't bear to be present when we botched it. But we hadn't. When he arrived late Sunday evening, we were sitting in *Adelphi*'s cockpit high above the darkened shop, drinking sodas, playing with the tiller, jabbering, drunken with success. We were as stunned as anyone that we'd gotten the job done, but we feigned nonchalance, as though we mounted a rudder a month and hadn't sweated blood over this one.

Doug examined our work from every angle, swung the rudder from side to side outside the vessel, climbed aboard and moved it with the tiller, and peeked inside the stern. Then he grinned. "I knew you could do it!" he said. "You just needed a guy like me to light a fire under you!"

And there you have the paradox of Doug White. The best I would have said about him back then was that his kids seemed to like him, which was not insignificant praise. According to his detractors, he was a rogue, a phony, a scoundrel, a blowhard, a bully, a con artist, a shallow conniver, and a superficial, pushy swine.

But Doug also got things done. He was smart and experienced enough to know that *Adelphi* would never be a profitable undertaking. No, she was a project he undertook because she struck his fancy. That also is high praise. In the era of fast food and strip malls, he stood out as an individual. Doug White was *interesting*.

⌐

About 50 people showed up for *Adelphi*'s launching **(Illustration 39)**. It was a circus and a zoo. There were families with kids, boatbuilding students, boatbuilders, dogs, couples from the yacht club, a few stuffed shirts in suits, and some random rubberneckers who stepped in off Ruston Way. There were coolers jammed with soft drinks, wine, and beer. At high tide, Doug's wife climbed onto some scaffolding and broke the traditional bottle of champagne with ribbons tied to it across the steel foot holding the bowsprit. It broke on the first whack, a good omen.

The White and Thrasher families rode *Adelphi* proudly down the ways. She looked great, with her gleaming white topsides, unblemished copper-red bottom, and rich mahogany cabin with a white top. In the water, her engine started with no grumbling, her propeller turned, and her rudder guided her off, as it should.

To the casual observer, it must have seemed a fairy-tale ending, the restoration accomplished, the participants smiling with good cheer and brother- and sisterhood. However, to a large extent, the launching was a magnificent sham, for *Adelphi* was only a pretty husk, a rind without the melon inside. Below decks she remained gutted, unfinished, lacking bulkheads, cabinetry, galley equipment, head, bunks, wiring, plumbing, or electronics. There also remained the expensive and laborious tasks of installing her deck hardware, stepping her mast, and rigging her. It would require as much expense and skilled effort to restore her completely to her original state as had already been expended.

I watched *Adelphi* until she disappeared around a point of land on her way to Tacoma Yacht Club. That was the last I ever saw or heard of her. As I turned away, I noticed a tall man standing ramrod straight at the edge of the sloping, rocky shore, his hands tucked nonchalantly into his pants pockets. He also had been gazing after *Adelphi*. He wore a light, plaid sports coat of fine-woven wool and a brown beret. Surrounded by people, he seemed aloof and alone.

It was Charles Udall. His countenance was aquiline, proud and fierce, but otherwise unreadable. When I said hello to him, he acknowledged my greeting only distantly, as though he didn't recognize me, which he probably didn't. I wondered what he felt, whether it was disdain at seeing Doug at the helm of his one-time nautical mistress, or longing for the brown-haired woman I'd seen in his photo album.

Chapter 14
Parting vignettes

Never sleep with a woman whose troubles are worse than yours.

—Nelson Algren, *A Walk on the Wild Side*

Soon after *Adelphi*'s launching, I received a call from my friend Irv Warner in Kodiak asking whether I was interested in a summer field job with the Alaska Department of Fish and Game. Irv was second in command of a massive project to study forage fish—primarily capelin, herring, and sand lance—around Kodiak Island. He offered me a position as a crew leader, which had a nice ring to it, except that the crew would be only myself and one other person. The job was to start in early May, a month hence, and three months before I would finish at Bates. It took me all of five seconds to decide.

"Yeah, I'll be there," I said.

It took me a lot longer to analyze why I made this decision. Leaving early would mean I wouldn't officially complete the Bates program, but that didn't seem important. Many boatbuilders around Puget Sound lacked pieces of paper stating they were boatbuilders; they'd learned on the job—Don and Chuck Knapp were cases in point. It wasn't that I felt I had no opportunities in boatbuilding, as none of Trumbly's students went long without work. Furthermore, a few weeks before, Ed Beck of Mantra Marine in Seattle had offered me a full-time job as soon as I graduated.

"It's hard to find people who can work with both wood and fiberglass," Ed had said to me.

Ed utilized his own experience as a longtime Alaska fisherman to design and build high-quality, shallow-draft, fiberglass seiners. He advertised his boats by fishing one of them around Kodiak. When other fishermen saw his boat riding low in the water, plugged with fish, they wanted one just like her, and Ed got more orders than he could fill. Mantra Marine was exactly the sort of job my training at Bates and experience at CLK Yacht Crafters had prepared me for. Ironically, maybe that was the basis for my immediate decision to return to field biology. It was a sad but unavoidable fact that production wooden boats were a thing of the past. If I worked at Mantra Marine, I'd have to live in Seattle and spend a lot of time laying fiberglass, a prospect about as attractive as chronic hemorrhoids.

I wonder if I'd have decided differently if someone had offered me a job building wooden boats. I'm not sure, but probably not. My feeble, inchoate rationalization for attending Bates in the first place was to build myself a cruising sailboat and disappear into the sunset. The main lesson I learned at Bates was that if all one wants is a cruising sailboat, by far the fastest and easiest way to obtain one is to find a high-paying job somewhere and then to buy one. It took Trumbly himself five years of evenings and weekends to build his 40-foot sailboat *Osage*, and he had the advantages of having a place to work, being able to afford materials, and knowing exactly what he was doing.

My foray into boatbuilding had been like hopping a fast-moving freight train, only to realize after 19 months that it was headed in a direction I wasn't sure I wanted to go. I'd taken the road less traveled and learned there were reasons why it was less traveled. Irv's offer provided a handy shortcut whereby I could skip back to the previous road I'd been on. I told myself that I'd simply take advantage of an all-expenses-paid camping trip in the Alaskan wilderness and decide at the end of the summer whether to return to boatbuilding.

I informed Trumbly I'd be leaving the program around the first of May, half expecting him to try to talk me out of it. I also asked him if I could spend my last month finishing my own sailboat. I felt especially bad because Patrick Chapman was going to leave around the same time as I **(Illustration 40)**. Trumbly fixed me with his Zen-master gaze.

"I'm sorry to hear you're leaving," he said, "especially since I need every experienced guy I can get right now to finish the T-38. But the decision is yours; you should you do what you feel you've got to do. As for your working on your own boat, I guess I owe you that, at least."

I felt guilty. Trumbly didn't owe me a thing; quite the reverse was true.

While I was in Tacoma, I wrote a number of articles for *National Fisherman* on topics related to boatbuilding. I don't know exactly how many, because I didn't save copies of the articles. Even now, I occasionally encounter one I've completely forgotten and have the opportunity to read it as though someone else wrote it—which, in existential terms, someone else did write.

In the course of my brief foray into nautical journalism, I'd corresponded with Bruce Cole, the West Coast editor of *National Fisherman*. Coincidentally with my leaving Bates in the spring of 1979, the publisher of *National Fisherman* intended to launch a new magazine called the *Small Boat Journal* to try to siphon off some of the spectacular success enjoyed by *WoodenBoat* magazine, first published in 1974. Dave Getchels, the managing editor of *National Fisherman*, had asked Joe Trumbly to write a regular column on lofting for the *Small Boat Journal*, with the first article to appear in the maiden issue.

Trumbly's column was to be called "Boats and Bevels," and the topic of his first article was frame bevels. Trumbly submitted the article on time, but a few weeks before the scheduled publication in March, Getchels informed Trumbly the *Small Boat Journal* wouldn't be running the column after all. Trumbly was hurt and mystified by this decision; in his world, a commitment made was a commitment honored.

Sometime during my last month at Bates, Trumbly invited me to accompany him to Seattle, where he had an appointment with the editors to discuss their reasons for canning his column. I was honored that he asked me along and welcomed the opportunity to finally meet Bruce Cole.

Cole, Getchels, Trumbly, and I had lunch at a restaurant in Ballard. Cole was a lanky man, charming and unaffected. He seemed honored to be dining with Trumbly, whom he obviously greatly respected. He encouraged me to continue submitting articles to *National Fisherman*. When he learned I was building a 14-foot sailboat, he remarked that he'd been looking for just such a day sailer and offered on the spot to buy my boat as soon as it was completed, sight unseen. I declined, but later wished I'd accepted his offer. The fate of my boat was to spend the summer on a pallet in the Port of Seattle, exposed to the sun, waiting for transport north by SeaLand. By the time it reached Kodiak in August, most of the exterior paint had peeled due to wax having been applied during the fiberglassing fiasco, and it looked like a derelict.

After lunch, Trumbly met with Cole and Getchels. He told me the gist of the meeting as we buzzed back toward Tacoma in his VW Bug.

"They thought my article was too technical," he told me, "and that it was beyond the comprehension of the average amateur. They said that the *Small Boat Journal* was intended to attract people to boatbuilding, rather than scare them away."

Trumbly seemed bitter. "I have a hard time understanding their attitude," he continued. "In my article, I drew on everything I've learned in 24 years of teaching frame bevels to beginners. I wrote it as simply as possible. The concepts just cannot be reduced any farther than I broke them down.

"The fact is, boatbuilding is not simple. I admit my article would require some thought on the part of the reader, and that's the problem. Hobby boatbuilders don't want to have to think. They want to jump in with both feet and build a boat like they might build a bookshelf. For them, it's easier to saw frames square and bevel them by hand than it is to learn to calculate the bevels—even though by cutting bevels directly, they'd save a day of hard work on a small boat.

"Getchels also didn't like my writing style," Trumbly concluded. "He thought it was 'overly folksy;' those were the words he used. He said he'd reconsider the column if I'd be willing to rewrite the article, but I told him I didn't have time for that."

Nonetheless, a few days later, Trumbly gave me a copy of his article and asked me to read it to see what I thought. The article did seem to have an unlettered flavor and an overly folksy style, and the voice shifted frequently from active to passive, personal to impersonal. In terms of organization, it jumped directly into how to make the master bevel board used as a tool in determining frame bevels, without first explaining what a bevel is, why frames must be beveled, and the ugly consequences of not beveling them. With these multiple defects, Trumbly's article would have failed as a finished essay in a first-year college writing class.

Unfortunately, I wasn't able at the time to pinpoint exactly what was wrong. Even had I been able to troubleshoot the article effectively, I lacked the pedagogical skill necessary to convey criticism in a constructive manner. When I attempted to explain what Getchels had meant by "overly folksy," Trumbly rapidly lost patience. I wish now I'd simply rewritten the article and told him, "Here's how I'd revise it; why don't you see what you think?" But even that might have backfired.

I kept a copy of the article. Rereading it now, I realize it is passably well written for someone who hasn't done much writing. Trumbly's folksy style does not now seem to me entirely inappropriate for the intended audience, and it could easily be toned down. Above all, I realize that a competent editor could whip the article into fine shape in an hour or so. The backbone and meat are there, and Trumbly included three masterfully drawn diagrams that aptly illustrate the concepts he explained in writing. All the article needs is a bit of introduction and some cosmetic work.

What I don't understand, then, is why neither Dave Getchels nor Bruce Cole undertook to edit the article. Both were knowledgeable about boats and supremely competent editors. It is true that no magazine editor can afford to extensively rewrite an author's submissions on a regular basis—this is not the editor's job. However, Trumbly was a quick study. If one of them had revised that one article to demonstrate what was expected, I suspect Trumbly could rapidly have become a competent writer. After all, writing a short article is child's play compared to building a big sailboat.

Of course, I have no idea what occurred at that meeting. Perhaps one of the editors did offer to rewrite it, and Trumbly refused, mistaking criticism of his writing for criticism of his knowledge of lofting. Perhaps he reacted the same way I had when the *Nor'westing* editor suggested I rewrite the *Adelphi* article. Perhaps neither editor had the time for a rewrite.

Whatever the case, the episode was tragic. Trumbly'd learned from masters—Adolph Cummings and Johnny Martinolich—in the heyday of production wooden boatbuilding. He'd soaked up their knowledge like a lamp wick and then spent a quarter century improving old techniques and inventing new ones. Arguably, he knew more about lofting boats than anyone else in the world. A slight nudge might have caused the boulder to roll in a different direction. It is easy to imagine that had the *Small Boat Journal* published Trumbly's first article, it would have spurred him to write more. In writing more, perhaps he'd have finished the book on lofting he often talked about.

The tragedy is that some of Trumbly's hard-gleaned knowledge will be lost, at least until someone independently rediscovers it. His innovations, now dispersed among those of his students still living and their students, will be diluted with the passage of time like lifeblood seeping into a river.

⌒

Armen Melkonian, Trumbly's de facto emissary to the East Coast, returned from his job at a boatyard in Maine. Things had not gone well, but we held a party in his honor anyway.

"They didn't pay very well, barely enough to live on." Armen said. "That wasn't the main problem, though. The main problem was that they had traditional ways of doing things, and they wanted me to do everything their way. When I tried to use techniques I learned from Trumbly, they acted like I was practicing witchcraft.

"I finally gave them a month's notice to find someone else. We parted in agreement—I was as glad to leave as they were to get rid of me."

⌇

Willie Hartman, the strongman from Kodiak who'd played car tag on the field trip to Evergreen College, returned to Alaska a month before I did. One evening after work, he'd stopped in a bar on the Flats and poured enough alcohol into himself to float a canoe. The story eventually reached us at Bates that, driving home, he not only ran most of the red lights on 11th Street across the Flats but also finally clipped a car and injured the old man driving it. Willie continued on his way with his Dodge Charger hemorrhaging parts, until several squad cars pulled him over. The story had it that Willie said to the first cop to reach him, "Good evening officer, is anything wrong?"

The cop was furious, thinking Willie was making fun of him. When he realized Willie really wasn't aware there'd been an accident, he calmed down a bit but threw Willie in jail anyway. Fortunately, the old man recovered. Willie spent his last couple of months in Tacoma without a car, embroiled in legal proceedings and insurance haggling. A judge revoked his driver's license and accepted a plea bargain whereby Willie could stay out of jail only if he returned to Alaska—hopefully forever.

It's difficult to find any humor in a case of aggravated drunk driving, but some did manage to seep out of the mess. Not long before Willie's departure, Patrick and I visited him at his apartment. He'd already sold his furniture, so we sat on packing boxes. Mortified by what he'd done to the old man, Willie didn't want to talk about the accident or its aftermath. All he could think about was getting out of Tacoma.

A knock interrupted our awkward conversation. At the door was a well-groomed, middle-aged guy in a suit. He was from a collection agency, come to pick up the Charger.

"Yeah, sure," said Willie. "I'm sorry I missed payments, but things have been pretty rough lately. Here're the keys. It's out in the yard." Willie handed him a ring with two keys on it.

About five minutes later, there was another knock at the door and the guy stuck his head back in. "Listen, Mr. Hartman," he said, "I can't seem to find the car. Could you come out and show me where it is?"

We all went out in front. Willie pointed to four wheels stacked together against a tree. "There she is, or what's left of her."

The guy gave Willie an exasperated look, but he was a true professional. He backed up a pickup truck, loaded the wheels, and departed. "We'll be in touch, Mr. Hartman," he said as he sped off.

Willie left Tacoma before learning whether two keys and four wheels legally constitute an automobile in the State of Washington.

⌐

Klarich had started dating Betty, a short, beautiful blond who was a student in the drafting program at Bates. In her late twenties, with a five-year-old son, she owned an old, two-story house a few blocks from the school. I asked Klarich how she could afford a house.

"She was married until her husband blew his brains out not too long ago," Klarich said.

"You moron," I said. "What's wrong with Kathy? And don't you wonder why Betty's husband blew his brains out? Maybe you'll be blowing your brains out in a couple of months, too."

Klarich was miffed. "Nothing's wrong with Kathy, except that we don't get along. And Betty's husband had serious problems; it wasn't anything to do with her. I love her. She has some shoreline property in the Narrows, and I'm going to help her build a cabin there."

To cement their relationship and drag it out of the closet, Klarich and Betty threw a big party one Saturday evening at Betty's house. Betty was a local girl and seemed to have countless friends, male and female. Klarich was also local, but the only people he invited were the boatbuilding class.

It was spring, and we were ready to cut loose after a long winter. The party should have been a mellow, convivial affair, but for some reason, it was not. There was an undercurrent of ugliness and frustration that surfaced as people drank. It might have been due to dissatisfaction among Betty's friends that she'd struck up an affair so soon after her husband's death, or dissatisfaction among Klarich's friends that he seemed to be thinking with the wrong head. Maybe it was just a misalignment of the planets.

Whatever the case, one of Betty's male friends, an automotive student at Bates, began badgering my boatbuilding classmate Joe Brezlin. "I don't like you," he said to Joe. "You look like a faggot to me, and you don't belong here. You'd better leave before I hurt you."

Joe looked uneasy but not afraid. He sat silently through minutes of abuse. If it'd been me, I simply would've left. I tried to avoid fights at any cost, including absolute humiliation. Thus I was astounded to find myself shouting at the guy that I thought he was scum; that by insulting another boatbuilder, he was insulting me; that if he didn't close his mouth, we could go outside and I'd do my best to pound it shut. I was angry beyond control and hoped with every atom of my existence that he'd head for the door.

The guy must have seen something scary in my eyes, because he was quiet for a few minutes. Then he attempted to save face by striking a bond with me. "Hey," he said, "I like a man who stands up for himself. You're a guy I could be friends with. I wanna shake your hand!" It was a coward's praise.

"Come on, Joe, let's get the hell out of here," I said, and we left. We spent the rest of the evening talking at his place.

"Listen," I said to him, "I was so far out of character wanting to fight that guy, I can't believe it. I hate fighting, and usually I'll do anything to avoid it. I don't know what came over me tonight."

"The curious thing is," Joe said, "I'm not afraid of fighting, but if I do it, I put my life at risk. I had a liver transplant when I was a kid. Now, if I get kicked in the abdomen or take a beating, I might reject my liver. For the same reason, I can't drink much, either."

I felt no pride in having stood up for Joe Brezlin. I'd humiliated him by doing so, because he'd been doing a superb job himself with Gandhian non-aggression, and I'd humiliated myself by allowing a bully to diminish me to his level.

Though I saw Chuck Knapp often socially, I visited CLK Yacht Crafters only once during the last couple of months I was in Tacoma. I was still a little miffed at getting laid off. In another attempt to bring in some fast cash, the yard had taken contracts to build three 32-foot, fiberglass Puget Sound gillnetters. They sat in the shop where Aquila's loft had been.

"We were making good progress," Griz told me, "until we ran into problems with the fuel tanks. We're going to have to yank them out, clean them, and re-install them. It looks like someone put sand in them."

"Who would do such a thing?" I asked rhetorically, thinking it had been random vandalism.

"I think it might have been the Indians," Griz answered, "but of course there's no proof."

182

To circumvent catch limits imposed by the Boldt Decision, some non-Native American fishermen had false holds or other hidden compartments built into their vessels. Griz reasoned that, if the Natives thought this was the case with the gillnetters at CLK, they might have carried the battle to the source rather than fighting poaching at sea.

∽

Several weeks before I intended to depart Tacoma, I informed one of my landlords I'd be leaving. I asked when I could get my $50 security deposit back.

"You can't leave," he said. "Youse gotta give us more notice than that. It says so in the contract."

I probably should have read the contract. Anyway, they wouldn't have any trouble at all renting the place, with its proximity to both Bates and the University of Puget Sound—provided they didn't let their nephew interview prospective tenants.

"I'm sorry," I said, "but I just learned yesterday of a job opportunity in Alaska, and I've absolutely got to leave by May first."

"Listen," the landlord said, "I'm gonna have to talk to my partners about this. I'll get back to you."

A couple of days later, he phoned me. "I talked to my partners, and we decided you gotta pay us the rent up to the end of your contract if you wanna leave."

I laughed at him, which probably wasn't smart. "You've got to be kidding. What are you going to do if I don't pay you, take me to court long-distance?" Later, I thought of various things a family of New Jersey hoods could do to me before I left.

"Listen," he said, ominously, "my brudder-in-law just got an associate degree in accounting up at Tacoma Community College, and if you wanna play hardball, we're gonna get him on your case."

When I realized he wasn't joking, I couldn't laugh any more. I was dealing with people who viewed a two-year degree at a community college as a serious source of power. In fact, the brother-in-law might have been the first person in his extended family even to finish high school, and I admired this. I might have been worried, too, if he had an associate degree as a paralegal—but accounting? It was just too sad. Nonetheless, I couldn't resist some academic one-upmanship.

"You can call your brother-in-law," I said, "but I have to warn you. I have a bachelor's degree in biology from the University of Alaska, and if he tries anything, I'll extract his otoliths." An otolith is a calcareous body in the inner ear of a fish; it has rings that can be counted to determine the fish's age. This was the only counter-threat I could think of on the spur of the moment, but for all the landlord knew, otoliths were an abstruse point of tenant law.

There was a long silence on the phone. "You prick," he finally said, unsure what manner of devious things I might be capable of with a bachelor's degree in biology. "One thing youse can count on, we're gonna keep your security deposit." He hung up with the sound peculiar to dial phones when the receiver is slammed down. That was the last I heard from my landlords.

⌒

One thing that mystified me was the relationship among Earl, Griz, and Breezie. When I first went to work at CLK, I noticed they often arrived at work together and left together. I assumed they just were close friends and lived in the same part of town.

Eventually Breezie and I became friends. This started one rainy day when I entered the lunchroom just as Earl was delivering her a humiliating tongue-lashing for some mistake she'd made. Breezie ran from the lunchroom in tears and slammed the office door downstairs as she left the building. From the lunchroom window I saw her walking away along Marine View Drive, her head down, getting soaked. Having myself been on the receiving end of some of Earl's masterful mind-humpings, I empathized and went after her. Breezie seemed glad to have company and spat out every foul epithet for Earl she could think of as we walked.

Not long after that, I asked Breezie if she'd like to have dinner together sometime. She accepted and suggested a fancy new rooftop restaurant at the north end of Pacific Avenue. I enjoyed the dinner, but curiously over the course of it, learned nothing of consequence about the person I was dining with. I found out she was from Oregon and had a mother, and that was it. As soon as I edged past the topics of the weather and the food, she put a wall up.

We went to dinner a couple of times after that at the same place, with the same result. Whenever I asked her to go anywhere other than to dinner, she declined. Finally I said one day,

"Breezie, you're a hard person to get to know. I'm wondering—why do you only want to go to dinner with me? I mean, there are other places we could go together."

She was quiet for a while, and then said, "Well, Earl told me that whenever a guy offers to buy me dinner, I should take advantage of a free meal. So I am."

"Oh." I said, as though her explanation made perfect sense. I couldn't get angry with her, because she didn't seem to be hitting on all cylinders. Neither did I, for that matter.

I didn't learn anything more until the following January, when Earl and I returned from Seattle on the last day of the boat show. We stopped by his place in Tacoma to unload some things from the show, and I was surprised to find Breezie there, looking at home. Griz and his girlfriend were there, too. It finally dawned on me that the reason Earl, Griz, and Breezie arrived at work together was because they lived together.

I then immediately jumped to the conclusion that they were living together as two couples. If this was the case, then it meant that I'd been hitting on Earl's woman and that he'd given her limited permission to be hit upon. On the other hand, I knew Earl went out with other women than Breezie. All of this was more than a simple country boy from Colorado could fathom.

One sunny Sunday afternoon in April, soon before I was to leave for Alaska, Earl invited Chuck and me to a combination housewarming barbecue and going-away party. He and his entourage had recently moved from Tacoma to a place in the countryside out past Puyallup. The house was situated in a clearing in the woods along a paved, two-lane road. Lilacs in the yard scented the air. The surrounding forest was tinged with the light green of newly budded leaves, and a chorus of frogs chirped in a nearby bog.

Earl's son Tim, nicknamed "Flash," was also there. He was a surprise to me, because Earl had never mentioned a son. Flash didn't live with Earl and the others; he'd driven from somewhere in a big sixties-model car with tail fins. In his late teens, he was lean and gawky, with brown hair conservatively cut. He wasn't what I'd have predicted in Earl's son. He was unobtrusive, polite, a little awkward, and eager to please. Flash and I threw a football for a while as the others talked.

Later over a plate of potato salad and chicken, I asked Earl about Flash.

"What can I tell you?" Earl said. "He's the most worthless kid you can imagine, no redeeming qualities. He's a complete pansy. If I'd known how he was going to turn out, I'd have strangled him in the crib."

Earl seemed serious, but if so, why would he have invited Flash to the barbecue? I jumped on him anyway.

"Why do you talk like that? You're his father, for chrissake. He seems like a decent kid to me, better mannered than most I've met his age."

"Well, time will tell." He grinned at me and changed the topic.

Later, while the others were cleaning up, Earl took me aside. "Everyone needs a family, right? I mean, people you can count to be there when you're in trouble, people who will watch your back without question. Can you agree with that?" he said.

Expecting him to tell me he'd been joking with his remarks about Flash, I said, "Sure."

"Well, a family doesn't have to be a biological family, right? Sometimes, people are closer to others they're not related to than to their blood relatives. Are you following me?"

"Yeah," I said. I could agree with that proposition.

"Well, Griz, Breezie, and I are a family. We're not biologically related, but we look out for one another. We work and live together, and share one another's successes and failures. Those two are closer to me than any of my biological family. Do you understand?"

"Sure," I said, "I think I understand."

"That's not all," Earl continued. "All the time I've known you, you've seemed sort of lost. You're alone, without anyone to look out for you. Griz, Breezie, and I talked it over last night; we all like you, and we want you to join us, if you want. You don't have to go back to Alaska. You can move out here, work wherever you want, do what you want, but you'll be part of a family."

Well, fuck all! That certainly wasn't where I thought Earl was headed with our little tete-a-tete! My ego ballooned at the implied praise, that Earl and company thought enough of me to single me out to join their exclusive club. At the same time, it deflated at the knowledge that they considered me so weak and aimless as to be suitable material for their mini-cult.

Relevant questions flitted through my mind like scared bats in a hallway. Who's the daddy in this family? Who's the mommy? Who does the spanking? What sort of person claims family values, yet badmouths his own son?

"You know, Earl," I said, "I'm honored that you guys invited me in, but to be honest, I don't think it would work out."

"Don't make a decision now," Earl said. "The offer's still open; think about it some more."

As I was climbing into Chuck's car to return to Tacoma, Breezie ran up and hugged me. She handed me a small photograph and said, "This is for you; don't show it to anyone."

The photograph was a 2 x 3 inch color head portrait of Breezie from her left front side. In it, her long neck emerges from a fuzzy white sweater; shoulder-length golden hair frames her long, pale face. She's gazing into the distance past the viewer's left shoulder. She's not smiling, and her expression is hard to read. There is sadness and vulnerability, yet also stubborn determination and a quality of endurance. On the back, she'd written the following in pink ink:

4-23-79—Matt, When this you see remember me & bear me in your mind. Let the world say what they may, speak of me as you find. Take care & God bless—Love, Breezie.

I no longer judge Earl's offer as harshly as I did in 1979. Whatever drew—or drove—Earl, Griz, and Breezie together back then must have been some heavyweight stuff.

⏤

I spent my last month at Bates working feverishly on my sailboat. I installed fir deck beams and covered the foredeck with plywood. I constructed a hollow wooden mast and a solid boom from Sitka Spruce. I built a spiffy centerboard from mahogany plywood with a cap of Alaska yellow cedar for contrast. I made a rudder from plywood and a stout tiller of laminated oak. I took the sail plan to a loft in Seattle and ordered the sail to be made. I coated the hull with several coats of white primer, sanding it between coats, and then applied two coats of canary-yellow enamel paint.

When the boat was as far along as it was going to be before I departed, I asked Trumbly to inspect it. He spent a minute or two looking it over.

"It's a pretty good job," he said. "It looks like you might have learned something here in Tacoma!"

From Trumbly, who'd have been quick to point out any serious flaws, that was a high compliment.

My last day in class, Trumbly handed me a piece of paper. It was a formal certificate with "L.H. Bates Vocational Technical Institute" printed at the top. On the next line was printed "Certificate of Completion," and below that, the usual gobbledygook dealing with authority, the board of trustees, dates, and so forth. The last line stated I'd completed 1,950 clock hours of training in the Boat Building Program.

The certificate is pint-sized, only two-thirds the size of a sheet of typing paper. It is a cheap form printed in black-and-white, with blanks to be filled in; there is no embossed gold seal or ribbon. My high-school diploma is four times as impressive, and compared to my college diploma the certificate is clearly a pauper among official documents. I've since accumulated even larger, fancier diplomas, including two written entirely in Latin from an ivy farm in the East. Still, if I ever hung my diplomas on a wall, and if someone pointed a pistol at my head and demanded I tear up all but one of them, that little scrap from Bates Vocational signed by Joe Trumbly is the one I'd keep.

Chapter 15
Trumbly's life

In the early 1900s, oil became part of the mix, and gusher wells in Osage County multiplied. There were only about 2,000 members of the tribe at the time. "We became the richest Indians in the world," said tribal council member Joe Trumbly. [A relative of the Joe Trumbly in this book.]

—Chris Welsh, *Star Tribune* (Minneapolis-St. Paul), 28 April 2002

Joseph Henry Trumbly was born in Elgin, Kansas, on 29 December 1917. He was the youngest of the 10 children of Clarence Augustus and Lillie Ola (née Todd) Trumbly, five of whom died in childbirth or at a young age. The surviving siblings were (in order) Gladys, Irwin, Norma, Francis (Fritz), and Joe.

In 1921, the family moved to Grants Pass, a town of around 4000 people in western Oregon. Clarence was 5/16 Osage Indian and had "head rights" in sharing the oil revenues that accrued to members of the Osage Tribe. His income from the oil money was enough that he didn't have to work very hard in Oregon, where he occupied himself with buying, renovating, and reselling property, and as a gentleman farmer. Through the Depression years, the Trumblys were locally regarded as wealthy.

I have no information on Joe Trumbly's early life, except that one of his daughters told me he had a "perfect childhood." Grants Pass is a beautiful place, occupying a valley surrounded by wooded mountains, with the Rogue River running through the center of town. In the 1920s, it was becoming a resort town, visited by the occasional movie star for its wild scenery and good fishing.

I imagine Joe's childhood was much like my own three decades later in Trinidad, a town of similar size and character in southern Colorado. In grade school, my best friend and I would travel the length and breadth of the town on bicycles, or hike for miles in the surrounding forests. Often on Saturdays, we would set out after breakfast and not return until dinnertime, and no one worried about our whereabouts. Society was safer then than now.

Trumbly's first exposure to boatbuilding came at age 12, in the 7th grade, when he and his brother Fritz constructed a rowboat from tongue-and-groove boards. One source stated that the boys' uncle Frank supervised this project and taught them the rudiments of boatbuilding, whereas Trumbly claimed that Fritz's 8th-grade friend Ferd Fleming was the driving force. The rowboat must have been a positive experience, because the following year Joe and Fritz built an 18-foot canoe framed with green willows.

Outside school, Trumbly's activities in his high-school years involved work and hobbies. Trumbly's first wife, who had known him in high school, told me he played football, but the high-school yearbook for his senior year shows no indication of this, nor indeed of any other extracurricular activities. Later, as an adult, he showed no interest in sports—when people tried to engage him in sports conversations, he'd joke that he got football and baseball mixed up. Nor did he show much interest in politics.

Curiously, the "Class Prophecy" prediction for his future, listed in the yearbook for each member of the senior class, is "Professor of English," which suggests he had a literary bent or at least stood out among his peers for literacy (unless the prophesies were intended to be ironic).

Trumbly got no allowance from his father and so was forced to work for any spending money he wanted. His first job, at age 13, was hand-digging 4-foot by 8-foot irrigation ditches on local farms. This was at the start of the Depression, when jobs were becoming hard to find, and there was some local resentment that Trumbly, whose family had sufficient income, was taking a job from someone who really needed it. Trumbly's response was simply to work harder than anyone else, to justify his hire. He spent his ditch-digging wages on model airplanes.

Subsequently he worked after school in a welding and machine shop, where he gained the skills that would eventually land him in a shipyard. On his own time, he used the shop to machine the parts for and build a functioning model steam engine, and to work on his four motorcycles, only one or two of which were running at any given time.

Trumbly graduated from high school in 1936 at age 18 and for the next four years moved around, taking whatever jobs he could find. Two days after graduation, he hitchhiked to Seattle and from there caught a boat to Alaska. It is unclear exactly where he traveled, but probably up and down the Inside Passage in Southeast Alaska between Ketchikan and Juneau. From various accounts, he worked as a cook, logger, truck driver, and deckhand, making several runs to and from Alaska. One source claimed that he also spent a year studying mechanical drawing at a technical college in Portland, Oregon, and if true, this could have been in 1937. In 1938, he was back in Grants Pass, again working in the welding shop, where he augmented the gas-welding skills he'd obtained in high school by learning arc welding on the job.

In 1940, with WWII underway in Europe, the United States was gearing up to assist its allies. Shipyards in Puget Sound needed welders, and Trumbly hired on at the Seattle-Tacoma Shipbuilding Company in Tacoma to work on C3-class cargo ships, many of which were later converted to Bogue-class escort aircraft carriers, also called "baby flat-tops." The same year, he married Jeanie Wetherell, who he'd been dating for several years—she'd been a high-school freshman the year he graduated. Trumbly worked on steel ships in Tacoma through the war, two years as a welder, two years as a lead man (leader of a work crew; a step below foreman), and finally as a welding instructor.

It was during this period that he became interested in sailboats. In his spare time, he began to frequent the Quaint Bookstore, reading as much about sailing as he could find. The bookstore's proprietor, who owned a 21-foot sailboat, took him sailing one day, and that cemented his interest. Enchanted with the idea of harnessing the wind for power and with the feeling of a boat under sail, Trumbly was hooked for life; wooden boats and especially sailboats became what some people might call an intense interest, and others call an obsession. Jeanie summed it up in an interview,

"After we moved to Tacoma, what we did every weekend, every day he had off from the shipyard, was drive around to look at boats being built, or boats in marinas. At home he always had a small boat under construction. We didn't have a garage, so he put up a canopy with four poles and a tarp next to the house. When he got home from the shipyard, he'd be outside working under lights late into the night. Mostly he built sailboats. I remember when we launched his first one; the thing would just swing around this way and that, and it made me uncomfortable. I finally realized that I really didn't like boats very much, and that I was afraid of the water."

When I mentioned the light epigram "Women and boats don't mix" that I'd heard Trumbly utter on several occasions, Jeanie replied, laughing, "Poor guy—yeah, I started that one, for sure!"

After the war, the shipyards geared down and laid workers off, and Trumbly did whatever he could to make ends meet. He worked for one brief period rebuilding engines at Titus Motors in Tacoma and another selling aluminum siding door to door. According to Jeanie, he also returned to Alaska and worked on a fishing boat.

In 1946, Joe and Fritz Trumbly used $3000 they'd inherited from their mother to start Trumbly Brothers Boatbuilders, which they set up in a vacant shop in Old Town, Tacoma. Their first boat was a runabout strip-planked with cedar. Their first customer was John Breskovich, owner of Pacific Boat Building in Tacoma. He paid $350 for the boat and had it delivered to his shipyard, but never used it. Later, when he hired the Trumbly brothers as lead men, he told them he'd just wanted to have their first boat.

Around this time, Jeanie and Joe Trumbly divorced by mutual consent, and Jeanie moved back to Grants Pass with their two daughters, Cheryl and Chris. The main problem appears to have been too great a disparity in interests.

"We never had any fights; we never had any trouble at all," Jeanie recalled. "We were good friends. People would ask me why, if we got along so well, we got divorced. The answer is that Joe and I just drifted apart; we became more like brother and sister. He was a good guy who got along with everybody. The only flaw in his character, if you can call it a flaw, was that he was so amazingly focused on his own interests. I thought he was selfish, I guess, but I realize now that he just couldn't get away from what he was attempting to do."

Breskovich contracted the Trumblys to build four Monk-designed tuna tenders and two 32-foot cruisers. The Trumblys built these boats cheaper than Breskovich could do in his own yard; in fact, they underbid the contracts and broke even or lost money, because they wanted the experience. After two years, Joe and Fritz dissolved their partnership and sold their equipment to pay off outstanding debts. Six months later, they went to work for Breskovich.

At Pacific Boat Building, the Trumbly brothers helped build fifteen 32-foot gillnetters their first winter and another 45 the following year. The pace was fast; the yard could produce almost two gillnetters a week.

Working with patterns for almost everything, crews of 10 men cutting pieces and 10 men installing them could frame a boat in 70 minutes and plank it in 7 hours. The lead man during the Trumblys' first year at Pacific was Johnny Martinolich, who Joe later regarded as the man he'd learned the most from.

When Martinolich left the following year, Joe Trumbly replaced him as lead man. Although the Martinolich-family boatbuilders had refined wooden boatbuilding to nearly an exact science during the preceding four decades, Trumbly soon made timesaving innovations. For example, when he arrived at Pacific, the planking crew was reading the bevels from each installed plank to be cut on the next plank up. Trumbly immediately devised a way, using a carpenter's square and standard bevel gauge, to read the bevels directly from the frames of a lined-off hull; the bevels could then be written on the plank patterns, and all planks cut to the correct bevel from the outset. Later he invented a specialized bevel gauge for this purpose **(Illustration 13b)**.

Trumbly had a falling out with John Breskovich around 1950 that led to a brief vacation from Pacific Boat Building. It was five minutes before noon one day. Trumbly's planking crew had the steam box full of frames, and he asked the men whether they wanted to start framing then or wait until after lunch. They opted for after lunch, and Trumbly went off use the bathroom. Breskovich happened by and asked where Trumbly was. Some of the slackers on the crew said they didn't know, even though he'd told them he'd be right back. When Trumbly returned, Breskovich confronted him accusatorily, asking why the crew wasn't framing.

Trumbly didn't like the tone and lack of trust, and quit on the spot. He drove straightaway to Tacoma Boatbuilding Company, known as "Tacoma Boat," and was immediately hired as a lead man. He worked there for five months and then took a job repairing boats for a cannery in Nushagak, on the Bering Sea coast in southwestern Alaska. After several months, he received a letter from Breskovich asking both him and Fritz to return to help build 25 more boats.

After finishing the boats for Breskovich, Trumbly decided to build a sailboat for himself, a 33-foot double ender of his own design but modeled after Colin Archer's designs. Like many boatbuilders, he had the idea of just taking off, maybe sailing around the world. He completed the hull in a shop next to a tavern toward the Narrows end of 6th Avenue in Tacoma, but a fire burned the shop and his boat to the ground. Trumbly thought someone from the tavern had probably tossed a lit cigarette into the shop, but of course there was no way to know. He was left with $1500 in insurance money, which might just have been enough to replace the tools and machinery he lost.

In the early 1950s, Trumbly worked for various boatyards in Tacoma, one of which was Puget Sound Boatbuilding, a Martinolich yard founded in 1942. Back again with his de facto mentor Johnny Martinolich, Trumbly was lead man on a crew building 30-foot gillnetters. To give some idea of the nature and pace of the work, I quote the following from an article I wrote for *National Fisherman* (Vol. 59, May 1978) based on interviews with Johnny Martinolich and Joe Trumbly.

"You can't believe the pace at which we worked in those days— there is nothing to compare with it now," says Johnny Martinolich. "There were no unions in the yards then, and a man could do any job which was needed: boat carpentry, welding, engine installation, pipefitting or electrical work. Many workers had a number of these skills.

"Now, a man can only do the job specified by his union. There was an enthusiasm back then to do the best job in the shortest possible time—there was a competitive atmosphere Much of that is gone now, when men simply wait for their eight hours to be up.

"I remember the time we constructed my own 30-foot gillnetter. The two Trumbly brothers, my brother Chuck, a couple of others and I completed the hull over the weekend. I had the stem, keel and transom ready to go on a Friday afternoon. On Saturday morning we set them up on the jig. I had the steam up, and we framed and planked her that day. By Sunday at 3:30 p.m. we had her caulked, painted and off the jig."

Joe Trumbly, who was lead man at the time, even resorted to trickery to get the most out of his men. "When you are handling steamed planks," he explains, "it is necessary to get them up immediately after you remove them from the box. If you don't, the steaming isn't effective. Most of the boats we built didn't require many steamed planks, but sometimes I had my men steam them anyway. They worked like demons to bend them hot, and they kept up the pace as long as planks kept coming out of the cooker. The psychological pressure induced them to work much faster than if they had been fastening the planks cold."

After the job at Puget Sound, Trumbly took a commission to build two 40-foot boats for the Coast and Geodetic Survey, and at the same time worked nights welding fuel tanks for Cummings Boat Company. He had completed both hulls and nearly finished one of the boats when, in March 1955, Tacoma Vocational School needed a boatbuilding instructor to replace Archer Dellplain, who was retiring. Verne Bates, Tacoma's vocational-technical director at the time, handpicked Trumbly for the position and offered him the job. Despite some trepidation about his ability to teach, Trumbly took the position. Not having much, if any, formal vocational training himself, he did what he knew; he ran the Bates program like a working boatyard, with him as foreman.

⌐⌐

I may have given the impression that while Trumbly professionally built any type of boat someone was willing to hire him for, his only interest off the job lay in sailboats, but this was not the case. Around 1952, not long after the fire that destroyed his double-ended sailboat, he happened to attend a powerboat race at Green Lake in Seattle. After seeing the one race, he decided that this was something he wanted to do.

Classes of racing powerboats are defined by the type (i.e., inboard or outboard) and size of engine, and the type and length of hull. Trumbly started out building and racing runabouts in the 36 Class, which required a stock (ordinary; not specially built for racing) Johnson outboard engine of 36 cubic inches displacement, but with few restrictions on the hull. In 1956 he switched to the more popular DU Class; DU stands for D utility, meaning that the boat must be a runabout, or single hull, rather than a hydrofoil; other specifications were a hull length of 13 feet and a stock Mercury 55H outboard engine of 40 cubic inches displacement.

From 1952 to 1965, Trumbly designed and built one or two racing boats a year, but he was a procrastinator. He'd take his time lofting, making patterns, and cutting out parts, and started building only a week or two before a race, often launching the boat on race day with the paint barely dry. He was one of the few racers who designed and built his own boats. He also designed his own propellers and had them cast locally, as the propeller was a major factor in winning a race.

Trumbly's bible for hull design was *Naval Architecture of Planing Hulls* by Lindsay Lord, a colorful character in his own right. With a PhD in naval architecture from MIT, Lord came to the notice of the US government during the Prohibition Era, when he took commissions to design high-speed, load-carrying boats for smugglers running rum from Cuba to the US. No law-enforcement vessels could catch Lord's designs. The government remembered Dr. Lord and during WWII gave him an officer's commission and unlimited resources to design planing hulls for the Navy.

Powerboat racing was and is not for the faint of heart. Races in the DU Class were typically five miles long, split into three, four, or five equal laps, with the final time being the average of the lap times. The races started like sailboat races, with the boats running fast on a plane at least a minute before the start and trying to cross the starting line at full speed when the gun went off. At speeds averaging 50 mph or more, dangerous collisions could occur at the start or during a race, and there was always the possibility of losing control and flipping the boat.

Trumbly liked to talk about powerboat racing more than any other aspect of his career and could recount many races in great detail. His first race was the 10-mile Foul Weather Race on the Puyallup River, in a 14-foot 36-Class runabout he'd completed the evening before the race. Most of the other boats were family fishing or water-ski boats, and Trumbly won the race like a wolf among poodles, with enough time to spare that he was on shore having coffee when the fastest of his competitors got to the finish.

Perhaps his most memorable race was the 1956 Sacramento River Race, 320 miles long from Sacramento to Redding, California. The race had several refueling stations, and his brother Fritz drove from one to the next to handle refueling. Sometime during the race Trumbly took the wrong channel and bent his propeller, but finally got back on course, replaced the propeller, and finished high.

Starting in the DU Class in 1956, he built and tested boats and propellers, and raced them hard, until he began to consistently win races. He did well in the Lawrence Lake race, considered by professionals to be one of the fastest courses in the world. He finally won the Sammamish Slough Race (Lake Washington to Sammamish Lake, Seattle) three years running from 1963 to 1965, and that was enough. In 1965 at age 48, he quit racing and moved on to other things.

Though Trumbly no doubt enjoyed the thrill and camaraderie of racing, I believe that above all he viewed racing boats as a quick way to experiment with designs. Within a period of a few weeks, he could draw a plan, build a boat, and test it to its limit. In fact, he could test boats faster than he could build them himself, because from 1955 to 1965 many powerboat racers took his class specifically to learn to build racing runabouts, and some of them built his designs. At any given time as many as five students in the class were building powerboats. On one occasion Trumbly took eight of his students' boats to a DU race, though it is not clear whether all of them raced, or whether he tested them all and raced the fastest one himself.

Trumbly's work in propeller design **(Illustration 15)** is a good example of his quickness in grasping concepts and improving on them. On one field trip the class took to Coolidge Propeller in Seattle, an old patternmaker explained how he laid out propellers on paper and then made the wooden models used to sand-cast them. Trumbly asked him a few questions about the principles involved and then went home and taught himself the whole process.

After designing many propellers and testing them on his powerboats, he became good enough that he was designing propellers for the top tier of professional powerboat racers. His propellers were used, for example, on several Miss Budweiser boats in the unlimited hydroplane class. Along the way, he invented a "pitch-o-meter" to measure the pitch of propellers and developed a folding sailboat propeller known as the Trumbly Tri-Blade, which he eventually patented. The idea behind folding propellers is that the blades collapse when not in use, reducing the propeller's drag on a boat under sail. Most folding propellers do not function well in reverse, and Trumbly's propeller addressed this problem.

~

Sometime after his divorce from Jeanie, Trumbly met and married a woman named Florence Orr and had another daughter, Susan. Jeanie related to me that, before Trumbly married Florence, he traveled to Grants Pass to inform Jeanie he was remarrying, and to ask for her understanding and blessing. I thought this was very unusual, and it indicates the level of friendship retained between Trumbly and his old sweetheart. In any case, blessing or not, the marriage to Florence did not last long.

In 1965, coincident with and perhaps causal of the end of his powerboat racing, Trumbly reconnected with and married Etta Royer, a woman he had worked with in the shipyard during WWII. Etta had been a welder on Joe's crew, and she and her former husband had lived close to Joe and Jeanie during the war.

Etta **(Illustration 41a)** was the exception to "women and boats don't mix" that Trumbly had probably been seeking his entire adult life. Joe and Etta bought wooded shoreline property on Raft Island west of Gig

Harbor and built a large A-frame home **(Illustration 41b)** on the side of a slope overlooking the water, with a ground-level shop where Joe could work on boats. They did the construction themselves, living in Tacoma at South Tyler and 19th Street until they'd completed the house. Etta had four children of her own, but most had already left home: Barbara, the youngest at 16 when Etta married Joe, was almost out the door, though she and her siblings came to know Joe well and liked him. When I was in Tacoma from 1977 to 1979, Joe and Etta shared at least one activity, which was participation in a bowling league one evening a week.

While still teaching at Bates and laboring on the Raft Island house, Trumbly began lofting and making patterns for *Osage*, a 40-foot ketch of his own design **(Illustration 19)**. When the house was finished, he devoted to *Osage* all the energy he'd previously spent on powerboats. After pushing hard in his free time for five years, Trumbly launched *Osage* down the slope from the A-frame to the water in August 1972. Double-planked with Port-Orford cedar over steam-bent oak frames, with a Honduras mahogany backbone and cabin, and a teak deck, all held together with 10,000 bronze screws, the boat was stronger and built with even more care than the hundreds of commercial boats Trumbly'd worked on. *Osage* was a masterpiece.

I don't know how much sailing he did, but he did not fulfill his youthful dream of a world cruise. He still had the Bates job, and—like many boatbuilders who build for themselves a craft capable of ocean cruising—he was more attracted to building than to cruising.

"Joe sailed *Osage* out to the coast once that I know of," John Possin (pronounced Poe-zeen´; a former student and later Trumbly's successor at Bates) told me. "Maybe he was headed farther south, but he got as far as Neah Bay, got becalmed, got seasick, turned around, and motored back." Possin chuckled about this.

Occupied with houses, powerboats, sailboats, day-teaching, and intermittent wives, Trumbly had his hands full, but one other professional activity took a chunk of his time. From his start as Bates instructor in 1955 until 1972, he taught the Marine Apprenticeship Program in Marine Carpentry one evening a week in the Bates shop. Apprentice carpenters at union shipyards (Carpenters Union, Local 470) such as Tacoma Boat could enroll in the program, which required that they put in 144 hours of class time a year. These hours counted toward the accumulated work time necessary to become a journeyman. Additionally, apprentices in the program received an upgrade, or pay raise, at their place of work every

198

six months. Some of the apprentices were graduates of the Bates program, but most were not. In 1972, John Possin took over as instructor for the Apprenticeship Program.

In the 1975–76 school year, Trumbly took a sabbatical from Bates, during which he traveled to boatyards, shipyards, and maritime museums on the Atlantic coast of the US and in Europe, studying historical wooden boats, talking to boatbuilders, and showing people the specialized tools he'd invented.

"Everywhere I went," he related, "I got to see loftsmen at work. They were good, don't get me wrong, but the Swedes were still transferring measurements with folding rulers. I showed them how to use marking staffs, but I don't think they understood a word I said. Anyway, it should have been obvious to them how much work marking staffs can save. Boatbuilders tend to be awfully conservative, though; if their grandfathers did something a certain way, then by god it's good enough for them. I don't know if they adopted marking staffs or not."

Aside from setting his Atlantic colleagues straight, the greatest effect of the sabbatical was on Trumbly himself. It was like a superbly accomplished provincial painter visiting the Louvre for the first time. The chance to observe construction details in historical ships from the Viking Era to the 19th Century put into perspective everything he'd learned over a period of 30 years. When I started at Bates in 1977, a year after Trumbly returned from his sabbatical, he was incorporating this historical perspective into his morning lectures.

My view of Trumbly's life from September 1977 to April 1979, when I was enrolled in his class, is covered in preceding chapters. I will, therefore, skip ahead to the years following his retirement from Bates. He officially retired at the end of July 1979 but stayed on for part of the following fall until his replacement could take over. The same fall, the Bates class launched the T-38, the 38-foot sloop I'd worked on; this was the boat Trumbly had sworn would go out of shop when he did, or we'd all die trying. The Bates administration set the minimum bid for the vessel at $46,500, which was the cost of materials plus 10%. Needless to say, the craftsmanship was fully professional—Trumbly did not let substandard work get by—and the boat was a good deal for whomever bought it.

Probably by 1977, Trumbly had designed the next boat he would build, a 51-foot, fin keeled cutter with a cold-molded hull, to be called *Windance*

(**Illustration 42**). This was to be his retirement project. By summer 1978, he'd completed lofting the boat, shaped all the backbone timbers, and built the transom. As mentioned, the boatbuilding class took a field trip to Raft Island that summer to set up the backbone (**Illustration 24**). By March 1979 he'd competed the frame molds and attached them to the backbone (**Illustration 43**). He'd also bored the propeller shaft hole and installed the engine, which is often done early in the construction of a boat due to the difficulty of maneuvering the engine into place later.

Trumbly stayed active in retirement, the main difference from the rest of his career being that he could work on his own schedule and on whatever he wanted. During one or more years, he taught a lofting workshop at the annual Port Townsend Wooden Boat Festival. In the early 1980s, he thought about designing and overseeing the construction of a boat that he would race in the BOC Challenge (a single-handed, round-the-world sailing race run in stages) but was unable to find a sponsor and had to drop the idea. He also contracted to help loft *Niagara*, a 11-foot brig from the War of 1812, for a complete rebuilding; the replica was launched in September 1988. During this period, he continued to improve and market his folding propeller.

Most of his time, however, went into *Windance*. It took him 10 years to complete the vessel, which was launched in summer 1988 and, after rigging, was ready for sea trials in summer 1989. This was undoubtedly a bittersweet time for Trumbly, for Etta's death from Lou Gehrig's disease in August 1987 overshadowed completion of the biggest boat of his life. As far as I know, this was the last boat he built.

After Etta's death and the launching of *Windance*, Trumbly seemed to gradually wind down. He remained in the Raft Island house for another 10 years, during which he apparently dabbled in boat and propeller design. He maintained connections with his daughters and stepdaughters, but he must have been very lonely without Etta. Around 1999 he sold the Raft Island property and moved to an apartment in Gig Harbor. A 1999 *Peninsula Gateway* article about Trumbly in Gig Harbor at age 81 indicates he was still driving a vintage Volkswagen Bug (the same model he drove when I was at Bates) and spent his time hobnobbing with friends at what he called "people watering holes" in Gig Harbor and Tacoma. The article is a summary of his life from direct interviews, but some of the facts are so skewed as to suggest that either the reporter was sloppy or Trumbly's memory was becoming a bit garbled.

Around 2000, Trumbly had an automobile accident that resulted in the revocation of his license. He concurrently began to exhibit symptoms of senility, with memory loss and failure to recognize people he knew or where he was. He spent time in a nursing home but was later under the assisted care of an around-the-clock housekeeper in his apartment, which was the state of affairs in 2003 when I had hoped to interview him. Joe Trumbly passed away peacefully in his sleep on 17 September 2004 at his home in Gig Harbor. He was 86.

⌒

Was Joe Trumbly a genius, as some people have referred to him? There are various definitions of genius, ranging from "someone with exceptionally high IQ" to "a person of extraordinary intellect and talent." Trumbly was intelligent, no doubt about that, but probably no more so than some of his peers, like Johnny Martinolich. What set him apart were single-mindedness of purpose, an intellect prepared by hard training, curiosity, and innovativeness. All boatbuilders faced the same problems he did, but most of them were content to use the techniques they'd learned, overcoming laborious problems simply by applying more labor. Trumbly encountered these problems and found novel, timesaving ways around them.

This was especially true with lofting. I once heard him say that he thought about lofting all the time—on his daily drives to and from Bates, in the shower, and in his dreams. For him, lofting became almost an abstract exercise, where he delineated problems other people weren't even aware existed and attempted to find solutions to them. Sometimes in lecture, to keep us on our toes, he'd introduce esoteric topics, and we'd drag out the lines drawing for his little round-bottomed boat to do something new. It was basically lofting, since a lines drawing is equivalent to a miniature loft.

"Today, I'm going to show you how to loft a cant-frame," he'd say, "one that is tilted forward near the bow, so that it requires no bevel."

Or, "Today I'm going to teach you how to loft any plane through the boat, should you ever need to do such a thing. So, take a ruler and draw some arbitrary slanted line through the profile plan."

Trumbly claimed to have invented some of the techniques he taught, and there is no reason to doubt him, because he generally gave credit where credit was due. He acknowledged, for example, that Adolph Cummings of Cummings Boat Company in Tacoma had taught him how to determine frame bevels directly from the loft. That was in 1947, two years before I was born, which meant Trumbly had been lofting for at least 30 years when I met him.

He talked often of writing a book on lofting and maybe he started, but unfortunately he either never finished it or never published it. The year after Trumbly retired, Allan H. Vaitses published a book entitled *Lofting*, which is useful but lacks some of the advances Trumbly taught at Bates. Had Trumbly published a lofting book with all he knew in it, and published annotated books of his designs, I believe he could have become as well known as accomplished boatbuilder-writers like John Gardner, Howard Chapelle, and Francis Herreshoff. Ultimately, however, he was a boatbuilder rather than a writer.

Joe Trumbly was undoubtedly "a person of extraordinary intellect and talent" in boatbuilding. I don't know whether this qualifies him as a genius, and to me the label just doesn't seem important. It probably wouldn't have to him either.

Chapter 16
Afterward

People

I designed and completed only one boat after I left Bates. In fall 1979, Mary Ann and I were together again and took a job managing the community store in the fishing village of Ouzinkie, Alaska, on Spruce Island near Kodiak. In the second-story warehouse of the store building, I built a 19-foot, pram-bow workboat I'd designed and launched it in spring 1980 **(Illustration 44)**. I named the boat *New Valaam*, the name given to Spruce Island by Father Herman, the Russian Orthodox monk who founded a mission at the east end of the island around 1810.

The inhabitants of Ouzinkie, not lacking wit, referred to my boat as "New Valium." I built this boat for $1500 in materials, used it hard year-around for six and a half years, caught $9000 worth of halibut with it, and sold it used for $1500. It served its purpose well, although the design was a bit shaky: in certain sea conditions the pram bow would plow into a wave, stop the boat dead, and dump passengers into the bilge.

After *New Valaam*, I occasionally did boat carpentry, but not much. In the summers of 1981–83, I crewed on the salmon-seiner *Faith*, owned by Zack Chichenoff of Ouzinkie. In fishing, it is standard practice that crew members are paid a percentage of the net value of the catch at the end of the season, and it is understood that they will help overhaul the boat and gear before the season. *Faith* was almost identical to the gitney *Devil's Paw* I'd renovated in Kodiak in 1976–77, described in Chapter 2. When I first worked for Zack, I had almost no seining experience, and the main reason he hired me initially was so I could repair *Faith*, which I did for a week or two at the start of each salmon season.

I fished tanner crab with Chris Opheim on *Lisa Ellen* in winter 1981. After a cold, brutal day hauling crab pots, we were anchored in Kitoi Bay, Afognak Island, and I was lying in my bunk reading a book about Albert

Einstein. I realized that I did not like fishing very much; that I might not live very long doing it, as roughly 30 fishermen died around Kodiak every year; and that Einstein's academic lifestyle was more appealing. I decided to enter graduate school in biology, which I did the following September.

For several semesters at Kodiak Community College in the mid-1980s, I taught a course called Introduction to Boatbuilding, in which students learned to loft and build a small plywood skiff. During this course, I was again reminded how Trumbly's teaching philosophy differed from the norm. One semester, my students had finished installing a skiff's chine logs, which required a considerable bend and needed to be soaked before bending. They got the logs on, but by the next class, one of them had snapped in two upon drying.

When Mark White, the full-time carpentry instructor at the college happened to stop by, he remarked, "What a bummer that it broke, when you're trying to make a favorable impression on the students. It's not a good experience for them, when things don't go right."

I felt entirely differently. The broken chine was a golden opportunity to teach one of Trumbly's key principles: any idiot can do something right the first time, whereas it takes a real genius to correct mistakes. This wasn't even necessarily a mistake; wood is unpredictable and doesn't always do what you want. If a problem arises, you correct it and get on with the job. The students carefully removed the broken log, cleared dried glue out of the notches it lay in, and successfully bent in another one.

In 2003 I repaired a small wooden sailboat given me by Chuck Knapp, my boss at CLK Yacht Crafters years before, whom I'd tracked down for an interview in Boulder, Colorado. With a half-detached transom, the boat was starting to rot in his backyard; along with the boat, he gave me a trailer to tow it away with. The same year, I helped my daughter Sarah build a 14-foot sailboat hull for her high-school senior project. Everything I'd learned at Bates a quarter century before was still in my head and hands; I was a little rusty, but I could still build a boat.

⌐⟍

CLK Yacht Crafters went out of business in 1980. To my knowledge, only one hull was ever cast from the Aquila mold. Earl Edwards ended up owning this hull, which he finished in 1983 and named *Imagine*. I later obtained a picture showing the boat in San Diego Harbor, and it was a beautiful vessel indeed. *Imagine* was still listed in 2013 in the boat registry for Kauai County, Hawaii, as owned by Earl George Edwards.

Chuck Knapp visited me for a few weeks on Spruce Island in the winter of 1980 and helped me finish *New Valaam*. He returned the following summer, and we used the boat to commercially fish halibut around Spruce Island. Chuck liked Alaska and stayed for two and a half years. He fished two years with Ouzinkie captain Theodore Squartsoff on *Ironsides*.

In the fall of 1980, Chuck helped Mary Ann and me build a cabin at Sunny Cove on land we'd bought from boatbuilder Ed Opheim, who lived at nearby Pleasant Harbor. The following year, Chuck and his wife Christie bought land and built a cabin near ours. In 1982, the Knapps left Alaska for Boulder, Colorado, where they studied Buddhism and attended Naropa University. Chuck eventually became a licensed clinical psychologist and remained in Boulder.

In 1981, when Chuck and I were both struggling to make land payments to Ed Opheim, Ed suggested we work off our debts by building him a 32-foot wooden salmon seiner. He would provide materials, tools, and a shop, and we would provide skilled labor. We might have agreed, but Ed's wife Anna was against the idea, and that put an end to it. I wonder how that would've turned out.

⌒

Several graduates of Bates Boat Building resided in Kodiak in the early 1980s. Willie Hartman had returned there, where he was raised, and was working at Ken's Boat Repair. In the mid-1980s, I'd occasionally spend an evening at the Village Bar, the establishment most frequented by Native Aleut-Eskimo fishermen, where I could expect to meet friends from Ouzinkie and catch up on local news. At the Village, I'd usually run into Willie Hartman. One night I'd had a few beers too many and was staggering toward the door, when Willie came up and took me by the shoulders.

"Give me your car keys," he said. "If you want to go home, I'll take you, but you're not going to drive. You're a professor now, and you make us Bates graduates proud."

I pointed out that he was as drunk as I was.

"I'm only a boatyard grunt." he said, "No one cares if anything happens to me, but you've got a family and other people depending on you." And he took me home. Although I wasn't in any condition to argue the relative value of boatbuilders versus so-called professors, I was deeply touched by his concern.

I left Kodiak in 1988 and did not see Willie again until I visited in

2000. At that time, he was working as a longshoreman in the Port of Kodiak. He declined my invitation to get together for a beer, so he may have quit drinking altogether. I forgot to ask him whether four wheels and a key legally constitute an automobile in Washington State.

After Bates, Patrick Chapman also returned to Kodiak and worked for some time at Ken's Boat Repair. Later, he helped Ray Tufts build a plywood 52-foot halibut boat they intended to fish together. Strangely, the close bond we'd had in Tacoma was gone, and we didn't see much of one another.

Like me, Patrick eventually left boatbuilding for another pursuit. His parents had started a photo-finishing business in Oregon, and when his father died, Patrick took over the business. When his mother died in the late 1990s, he moved to Hawaii to rediscover his roots. In 2013, after release of the Kindle edition of this book, I spoke with Patrick by phone. By then, he was going by the name Patrick Kanahele and living in Captain Cook, Hawaii, where he worked full time as the custom carpenter for a private estate.

↬

George Chambers, the tacitly acknowledged star in my cohort at Bates, made good use of his training. He bought a design from Trumbly for a 33-foot sailboat, which he built in the early 1980s and lived on with his wife Ruby for some years. In the same period, there was a mini-boom in constructing wooden luxury yachts for Microsoft millionaires in the Puget Sound area, and George worked on those. After that, he and Ruby started a general construction company that specialized in renovating old buildings. Ruby, who'd studied upholstery at Bates, went on to become an internationally recognized designer of custom-made bedding.

As of 2013, George and Ruby were major property owners in areas undergoing urban renewal in downtown Tacoma. They were involved in Spaceworks Tacoma, a joint effort between the City and the Chamber of Commerce to make no- and low-cost temporary space available to artists, artisans, organizations, and community groups—the idea being to draw people to the areas undergoing urban renewal. George and Ruby are a good example why money spent on public education is money well spent: the tax revenues they generated indirectly or directly through their enterprises likely repaid the Tacoma School District many times over for making available their training at Bates.

↬

I visited Mark Klarich in North Tacoma around 1982. We walked from his apartment to a rented garage nearby where he'd built a traditional-style lapstrake rowing skiff of his own design. The boat was so beautiful and finely crafted that it took my breath away. With a clear finish accentuating the fine woods in it, the boat looked like the product of a luthier rather than a boatbuilder, and like it belonged in a museum rather than in the water.

I was later shocked to learn that Mark died in a 1998 boating accident on Harts Lake near McKenna, Washington. According to a newspaper report, he'd been fishing from the shore but accepted an invitation to join two other men in an 8-foot boat. The boat capsized when one of the men stood up to take a leak; two drowned, one of whom was Mark. Investigators suspected that alcohol was involved, though relatives maintained that Mark hadn't touched a drop in years. Knowing alcohol and knowing Mark, he could've been dry for years but stepped off the wagon and onto the boat that day. Whatever the case, I valued his friendship and miss him.

⌒

Every story needs a villain, and Earl Edwards serves as a mild villain in this story. While Earl may have figured in the demise of CLK Yacht Crafters and emerged from the fiasco seemingly in better shape than anyone, far more powerful forces were involved, such as the familial tension between Chuck and Don, and Chuck's desire to quit boatbuilding. Ironically, among all of us at CLK, Earl was the only one to continue boatbuilding after Tacoma.

When I worked for CLK, I designed a logo (a group of stylized wind gusts) for the Aquila line. In 2000 I received an email from Earl telling me he was living near the beach on Kauai, Hawaii, and building a 118-foot, ultra-light, ocean-racing trimaran named *Rave*. He asked whether I could design a logo for that boat as well. I declined but was glad to hear from him. I was also saddened to learn that Griz, Earl's sidekick and my co-worker at CLK, had drowned in Hawaii. Earl passed away at age 73 at Niumalu, Lihui (Kauai) on 14 April 2014.

⌒

Doug White remains an enigma. Unable to find information on him or *Adelphi* in the Tacoma Public Library Northwest Room, I finally explained to a reference librarian that I was looking for a man who'd been a former Tacoma city manager. The librarian made short work of that; after consulting a list, he stated unequivocally, "There's never been a

Tacoma city manager named Doug White."

What he did find was an obituary for a 49-year-old Douglas E. White who'd passed away on 9 December 1984, five years after I left Tacoma. Survived by a wife named Judith and a son named Stephen, he'd been a member of Tacoma Yacht Club. It fit my Doug White to a T. This Doug White was a graduate of the University of Washington and the University of California, and was a Navy veteran of the Korean War. At the time of his death, he was manager of outpatient services at the Greater Lakes Mental Health Clinic.

I might have remembered incorrectly that Doug told me he was a former city manager; maybe he'd said "city engineer" or "city employee," but I don't think so. I remember thinking at the time, "Wow, I've been in Tacoma less than two years, and I'm associated with a former city manager!" An enigma in life, he remains an enigma in death, if I've got the right Doug White.

⌐◡

Ed Opheim passed away in February 2011, at age 100, in a nursing home in Kodiak. He was a remarkable man. From the 1960s through the 1980s, he was the best-known boatbuilder in Alaska. In the late 1970s, when he was in his late 60s, Ed bought a stock 40-foot fiberglass sailboat hull, named it *Ragna Olsen*, and began work on the interior and rigging. His goal was to revisit his old haunts from Kodiak westward. Though advancing age or lack of funds prevented him from finishing the project, it is yet another example that a common dream among boatbuilders is to sail off into the sunset.

Though he never went past high school, Ed was a self-taught scholar and writer. He spent winter evenings at Pleasant Harbor reading from his extensive library, and one question he grappled hard with was the existence of God. He'd discuss this question with anyone who stopped by and was well prepared to do so, because he'd read the Bible, Koran, and Book of Mormon from cover to cover.

Ed published articles and stories in *Alaska Magazine*, and three books (*Old Mike of Monks Lagoon*; *The Day the Meadowlark Sang*; *The Memoirs and Saga of a Cod Fisherman's Son*). The *Memoirs* is an intriguing account of Ed's early life in western Alaska, Montana, Seattle, and the Kodiak area; though it is almost impossible to find, I highly recommend it.

When I left Spruce Island for graduate school in August 1981, I asked Ed to keep for me a portable metal file box containing the negatives and proof sheets for over 700 black-and-white photographs I'd taken while in boatbuilding school. His house was heated in winter, and I didn't want to leave them in my unheated cabin. When I returned to Pleasant Harbor the following spring to retrieve the file, Ed didn't remember anything about it. Since I'd left in a hurry, I thought maybe I'd left it with someone else instead, but no one remembered it.

The photographs stayed lost for 17 years, until I arrived home one day in March 1998 in Middlebury, Vermont, to find an unexpected parcel sitting on my doorstep. Cleaning out one of the old laborer's shacks at Pleasant Harbor, Ed's daughter-in-law Kathy (Putney) Opheim had found the file and mailed it to me.

I still don't know how the file box got from Ed's place to the shack, or why Ed didn't remember having it. Whatever the explanation, I would not have attempted this book without the photographs, which constituted an indispensable graphic diary for reconstructing the time line of my stay in Tacoma and for jogging my memory.

↬

Bates Boat Building after Trumbly

Trumbly formally retired as instructor at Bates at the end of July 1979 but stayed on the following fall for a month or two until his successor was able to start. He handpicked as his successor John Possin, a former student who'd completed the Bates program in December 1959. Possin was well qualified; after graduation, he worked at Tacoma Boat for 20 years until he took the instructorship at Bates. He overlapped with Johnny Martinolich at Tacoma Boat for about five years and, coincidentally, worked there with Trumbly's brother Fritz.

Possin started out his career building wooden hulls, but in the 1960s and 1970s Tacoma Boat shifted to steel hulls with wooden decks (torpedo retrievers and a minesweeper) and then to all-steel decks and cabins (tuna boats, oil barges, crab boats). After the transition to steel, Possin built bulkheads and did finishing carpentry. Kiko Johnson of Hilo, Hawaii, who finished the Bates program under Possin, told me that while Possin endeavored to teach what was standard in the industry at the time, and thus focused more on fiberglass and aluminum than Trumbly had, he was an excellent, knowledgeable instructor.

↬

Possin retired from Bates in 1998 and was succeeded in September that year by Mike Vlahovich, another of Trumbly's former students. I'd met him working on the *Adelphi* restoration project in 1979, described in Chapters 12 and 13. Vlahovich lasted less than a year as instructor. According to Possin, Vlahovich had an opportunity to go to Alaska in June 1999 and asked for a leave of absence, but apparently did not approach the Bates administration in the right way.

"If Mike had approached me and asked me to teach the class as a substitute for a few weeks," Possin told me, "it could have been arranged, but he got huffy about it and either quit or got fired."

Possin thus ended up filling in anyway until the end of the summer quarter. Vlahovich's contributions lay in other directions than as Bates instructor. He'd already become a major player in restoring traditional wooden boats and preserving the history of boatbuilding, and he continued in this direction. He co-founded the Working Waterfront Museum in Tacoma in 1994; worked at the Chesapeake Bay Maritime Museum as director of special projects and boatyard manager; received the Washington State Governor's Art and Heritage Award in 1999 for his preservation of boatbuilding heritage; and was later founding director of the Coastal Heritage Alliance.

The next—and as it turned out, the last—instructor was Chuck Craydon, starting in September 1999. Chuck had a background in boat finishing carpentry from Northcoast Yachts and Delta Marine in Seattle, both of which built large luxury yachts. He also had academic credentials as an instructor of boatbuilding and sailing in a vocational program in Bellingham for at-risk youth. I visited Chuck in Tacoma in 2003 and 2004. By that time, LH Bates Vocational Technical Institute had been renamed Bates Technical College, and the boatbuilding program had moved to the South Campus, a spacious new facility in Tacoma's South End **(Illustration 45)**.

The new shop and classroom were several times as large as those on the Downtown Campus where I'd trained a quarter century before. The size of the shop was astounding. In one corner sat a sister boat to the 24-foot sailboat *Seraffyn* in which Lin and Larry Pardey had circumnavigated the globe in the early 1970s; the shop dwarfed the boat—it could have held 50 *Seraffyn*s. Many of the large power tools had been transferred from the old shop, but there were new machines as well, like a big horizontal sander. Adjoining the main shop were a separate paint room and aluminum shop, and there was also a loft. Scattered among the power tools on the shop floor were about 10 boats of various sizes, in various stages of repair or construction.

A big difference from Trumbly's class was that a few women were now in the program. Chuck Graydon's star student was a young woman named Christina Price, an Army veteran taking advantage of the GI Bill to get training. Christina's husband, a retired career Army man, was in the same class; the Prices' time-honored goal was to build a cruising sailboat. I asked Christina about her military service.

"I spent most of it in Korea," she answered. "I was in Ordnance and my job was testing. I spent every day blowing shit up and every evening drinking. It was fun."

I encountered two old friends in the new Bates shop. One was an unfinished 19-foot, fin-keeled, bent-framed, carvel planked sloop **(Illustration 9a)** that a student named Brian Saucier had been building at Bates during his last year, which was my first year. Trumbly referred to this boat as the "biggest small boat in the world," because although it was in the size range of a skiff, its construction was that of a much larger boat. Some of the beginners at Bates helped Brian set up the backbone and bend the frames, and we were able to observe techniques we would not otherwise have seen. Brian left Bates before planking the boat and never completed it. He'd recently donated it to the boatbuilding program as a project for someone to finish.

The other old friend was a seven-and-a-half-foot boat built from the lines Trumbly had drawn for his lofting exercise. Some of Graydon's students had undertaken this project, and the actualization of the lines sat on a strongback in the loft, framed but not planked **(Illustration 46)**.

Aside from the large size of the new shop, what struck me most about the Bates program in 2003–04 was the lack of energy. On the days I visited, I saw no more than seven students in the shop at one time, all of whom seemed to be working lackadaisically on small projects. In Trumbly's day, there would have been 15 students working like their lives depended on it, hurrying to and fro, generating sawdust as fast as they could. I asked Chuck why he didn't have his students working on a large project.

"The school is shying away from big projects now, after having gotten burned," he said. "John Possin began that sister boat to *Seraffyn* in 1984; the hull and interior are finished, but it needs rigging. It has the finest materials—fir backbone, mahogany planking, teak decks, bronze and stainless steel fastenings—and first-rate craftsmanship. The materials alone cost $15,000 at 1980s prices and would cost twice that today. It's been sold now, but the highest bid was only $9000. It was an absolute steal at that price, and the school took a big loss.

"But that isn't the real problem," Chuck continued. "The real problem is lack of motivation in the students I get now. It's different than before. There's no lack of projects right now in the shop, boats waiting to be finished or restored, but the students seem to feel that since they're paying for their training, they have the right to work on whatever they want, rather than on what I assign them."

Another huge difference from Trumbly's time was in the organization of the program itself. Trumbly taught on the apprenticeship model, where instruction was measured in clock hours, that is, the time a student engaged in supervised work related to boatbuilding. Trumbly took clock hours seriously; if he saw you standing around shooting the bull for more than five minutes, he'd assign you something to do.

In contrast, Bates was now labeled a technical college rather than a vocational school. Students worked toward a Certificate of Training or—if they also took courses in math, English, and human relations—a 2-year Associate of Technology degree, both of which were measured in terms of course credits rather than clock hours. Knowledge was parceled out in discrete units: BOAT 091, Occupational Human Relations; BOAT 113, Boat Design III (Lines); BOAT 114, Boat Design IV (Lofting); BOAT 121, Patterns; BOAT 203, Wooden Boat Tool Maintenance; and so forth.

A Student Program Tracking Manual included 31 checklists for students to keep track of what they learned. A thick guide explained "Competency Based Education" ad nauseam, and contained course goals and more checklists to be signed by the instructor.

In my opinion, the new system might well have shifted students' focus in training from working as hard and learning as much about everything as they possibly could, to completing checklists and obtaining credits with as little work as possible.

I can think of several reasons why Bates changed from a vocational-technical school to a college, with corresponding changes in the structure of training, as follows.

Public perception. In the 1970s, a college degree imparted greater status to the bearer than a vocational certificate, and this may be even truer now.

Transferability. People want verification of training they can take anywhere, and an associate degree from an accredited college perhaps fulfills this requirement better than a vocational-school certificate.

Documentation and accountability. In a society trigger-happy for lawsuits, institutions increasingly need to cover their asses. In less than a decade, for example, three groups of students sued Bates Technical College, claiming the training they'd received had not prepared them for the job market. Courts awarded settlements of $1,500,000 to 15 former students of the Denturist Program in 2002, $170,000 to students in the Court Reporting Program in 2007, and $500,000 to 16 former students in the Civil Engineering Technician and Surveying Program in 2008. It was no wonder Chuck Graydon was bogged down in checklists.

⤺

Bates unexpectedly terminated its Boat Building program in 2012. After training students in this trade for 63 years, the school abruptly informed both the students and Chuck Graydon that the program would stop at the end of the ongoing quarter. Students with projects underway had to move them somewhere else, and Bates sold the shop equipment and remaining boats. I won't second-guess the administration's decision, but it probably resulted from a cost-benefit analysis showing that the number of students trained or the level of demand from industry no longer justified the expense of the course.

In fact, it is difficult to pin down any one factor for the decline of the Bates program. In the course of interviews for this book, some people blamed one instructor or another after Trumbly for letting the ball slip. For every detractor, however, there were also supporters. All three instructors after Trumbly were competent boatbuilders and probably capable teachers. John Possin completed the program under Trumbly, knew Trumbly's teaching methods, and in theory should have provided continuity.

Trumbly was an extraordinary teacher, which by definition means he was hard to replace. It is an open question, however, whether the Boat Building program would still be going strong if Bates had found a succession of teachers with Trumbly's enthusiasm and energy, because the world changed in many ways after he retired in 1979. In the 1980s, the Gulf of Alaska crab fisheries collapsed, the bottom fell out of the price for salmon, and consequently orders for new commercial fishing boats dried up. At the same time, the cost of fine woods skyrocketed, making large wooden boats prohibitively expensive.

The academic environment also changed. I doubt Trumbly would have functioned well in the college system of course credits and checklists that Bates eventually imposed. In addition, entirely new, massive industries emerged that demanded trained workers and competed for training resources. The 2021–22 Course Catalog for Bates Technical College lists at least 19 degree programs related to computer systems and repair, information technology, and software development. A job in a boatyard or shipyard involves hard manual labor in a rough environment, and unless a young person simply happens to like boats, the opportunity to sit in a clean, air-conditioned office or shop working with computers at probably a significantly higher wage is more attractive.

It was a shame that Bates closed the Boat Building program, with its illustrious history, as there is still a lot of boatbuilding going on in the Pacific Northwest and other parts of the US. Some long-established companies like Tacoma Boat and Martinac have disappeared, but in the Puget Sound region, Modutech, Nordlund, Delta Marine, and other, smaller companies are hanging on.

There is also a steady interest in building small wooden boats. *WoodenBoat* magazine's 2013 Directory of Boat Schools listed 25 programs in the US that taught aspects of the trade. While I cannot find a comparable directory in 2022, a quick Internet search found at least 14 wooden boatbuilding programs currently available in the US, and there are undoubtedly more. This is encouraging. Although Bates is no longer on the list, the profession will continue.

Bibliography

General references

Aird, Forbes. 1996. Fiberglass & Composite Materials: An Enthusiast's Guide to High Performance Non-Metallic Materials for Automotive Racing and Marine Use. HPBooks, New York.

Baader, Juan. 1962. The Sailing Yacht. Translated from the German by James and Ingeborg Moore. Norton, New York.

Berry, Don. 1963. To Build a Ship. Viking, New York.

Bradley, Cliff. 1946. Building the Small Boat. MacMillan, New York.

Chapelle, Howard I. 1941. Boatbuilding. Norton, New York.

Gardner, John. 1978. The Dory Book. International Marine, Camden, ME.

Grayson, Stan. 1981. The Dinghy Book. International Marine, Camden, ME.

Hankinson, Ken. 1974. How to Fiberglass Boats. Glen-L, Bellflower, CA.

Hoadley, R. Bruce. 1980. Understanding Wood. Taunton, Newtown, CT.

Kent, Rockwell. 1930. N by E. Random House, New York.

London, Jack. 1911. The Cruise of the Snark. MacMillan, New York.

Lord, Lindsay. 1963. Naval Architecture of Planing Hulls. 3rd Edition. Cornell Maritime, Cambridge, MD.

Opheim, Ed. 1994. The Memoirs and Saga of a Cod Fisherman's Son. Vantage, New York.

Pardey, Lin, and Larry Pardey. 1976. Cruising in Seraffyn. Seven Seas, Camden, ME.

Phillips-Birt, Douglas. 1979. The Building of Boats. 1st American Edition. Norton, New York.

Ransome, Arthur. 1927. Racundra's First Cruise. Jonathan Cape, London.

Slocum, Joshua. 1900. Sailing Alone Around the World. Century, New York.

Spectre, Peter H., and David Larkin. 1991. Wooden Ship. Houghton Mifflin, Boston.

Spurr, Daniel. 2000. Heart of Glass. International Marine/McGraw-Hill, Camden, ME.

Steward, Robert M. 1994. Boatbuilding Manual. 4th Edition. International Marine/McGraw-Hill, Camden, ME.

Vaitses, Allan H. Lofting. 1980. International Marine, Camden, ME.

Verney, Michael. 1973. Complete Amateur Boat Building. American Edition. MacMillan, New York.

Witt, Glen L. 1960. Inboard Motor Installations in Small Boats. Revised Edition. Glen-L, Bellflower, CA.

Articles related to Trumbly or Bates Boat Building
[Chronological order, with annotations in square brackets]

Garrison, Ed. 1954. Along Tacoma's Waterfront. *The Tacoma Sunday News Tribune and Ledger*, 21 March. [Account of a 38-foot, Monk-designed troller built in the Bates Boat Building program and about to be auctioned; information about the early days of Bates Boat Building and its first instructor, Archer Dellplain.]

Anonymous. 1954. 38-foot troller built by students of Tacoma Vocational School will be sold. *The Marine Digest*, 20 February. [About the same boat and auction as in the Garrison article above.]

Winkler, Pat. 1960. Along Tacoma's Waterfront. *The Tacoma News Tribune*, 26 June. [Account of an Alaskan Eskimo student enrolled in Trumbly's class.]

216

Ferguson, Howard. 1968. How to have fun while helping others. He enjoys teaching boat building trade. *The Tacoma News Tribune and Sunday Ledger*, 17 November. [General article about Trumbly and the Boat Building program.]

Sypher, Dick. 1972. Boatbuilder gets around to build one for himself. *The Tacoma News Tribune*, 26 August. [Brief account of Trumbly's 40-foot sailboat *Osage*.]

Dick, Matthew. 1978. Martinolich family helped father West Coast fleet. *National Fisherman*, May, 59(1): 52–53. [Brief history of the Martinolich family boatbuilding dynasty, based on interviews with Johnny Martinolich and one interview with Trumbly.]

Dick, Matthew. 1979. Stock hull lengthened 5' to produce fine charter vessel. *National Fisherman*, February, 1979: 68. [About *Macs' Effort*, Aquila, Knapp Boatbuilding, and Chuck Knapp and Earl Edwards.]

Lane, Bob. 1979. Bates shipwrights launch last Trumbly. *The Tacoma News Tribune*, 3 August: A-5. [About Trumbly's completion and launching of his last class project, a 38-foot sailboat, and his pending retirement from Bates.]

White, Mark. 1981. Ed Opheim's dories. *WoodenBoat*, No. 43 (November/December): 40–43. [Not about Trumbly or Bates, but relevant to the text.]

Jacobson, Joseph A. 1989. Joe Trumbly. A master at the tricks of the boatbuilding trade. *WoodenBoat*, No. 89 (July/August): 44–51. [A detailed, knowledgeable, well-illustrated summary of Trumbly's career and accomplishments, written by a boatbuilder from interviews with Trumbly. The article notes the launching of *Windance*, Trumbly's last boat.]

Lande, Ann Coval. 1999. Local 'character' busy making friends. *The Peninsula Gateway*, 22 September. [Brief, seemingly flawed sketch of Trumbly's life and accomplishments, and his social life in Gig Harbor at age 81.]

Anonymous. 2004. Joseph Trumbly. *The Tacoma News Tribune*, 26 June. [Trumbly's obituary.]

Glossary

acetone—In fiberglassing, a volatile organic solvent used to remove liquid polyester resin from tools, hands, clothing, and anything else; it is also the main component of nail-polish remover.

adze—A chopping tool with a sharp horizontal blade and a gently bicurved handle longer than a hammer handle; held with both hands, it is used to trim away wood.

apparent bevel—A measured bevel that is incorrect because the piece to be beveled will itself have an angled orientation in final placement.

backbone—The part of a wooden hull's skeleton comprising the heavy structural members in the midline, including the stem, forefoot, keel, horn timber, transom, and various knees and fillers linking these elements.

batten—A long, flexible piece of wood with straight edges, used to mark long lines on a loft or boat skeleton, or used for fairing, in which case it is called a "fairing batten."

beam—1) The width of a vessel at the widest point. 2) A transverse, usually slightly curved skeletal element supporting a deck (deck beam) or cabin top (cabin beam).

bevel—An angle on a piece of wood, measured perpendicular to the cut. See also: standing bevel, under bevel, real bevel, apparent bevel.

bevel gauge—A tool shaped like a folding pocketknife but with a flat, straight-edged blade; used to measure or mark an angle.

bevel stick—A flat piece of wood about 2 feet long and 3 inches wide, marked with angled lines in 1-degree increments from 0 degrees to about 60 degrees; used to read angles from a bevel gauge or to set the gauge to a particular angle.

bilge—The bottom of a boat, just above the keel; also, the strongly curved portion of a round-bottomed hull where the sides turn sharply toward the bottom or keel.

block plane—A small carpenter's plane held in one hand and used for fine trimming of wood surfaces or edges.

body plan—The part of a lines drawing that shows the outlines of transverse sections of the hull (the shapes of half-frames or half-molds) at regular intervals.

Boston batten—A homemade ruler used in the process of lining off a hull to determine the widths of planks in a sector at any point along the length of the hull.

bow—The front end of a boat.

breast hook—A horizontal, V-shaped, knee-like element at the bow of a boat, strengthening the joints where the sheer clamps meet the stem.

bulkhead—A vertical partition inside a boat; analogous to an interior wall in a house.

bulwark—The extension of the side of a vessel above deck level, comprising a barrier to keep people and things from washing overboard.

camber—The convex lateral curvature of a boat's deck or cabin top.

cant frame—A frame member near the bow that is angled forward so as to lie flush with the planking without needing an appreciable bevel.

carlin (carling)—A long longitudinal member fastened along the insides of the deck beams to help strengthen the skeleton of a boat.

carvel—A type of planking in which the planks are laid edge to edge, with a V-shaped groove between them to hold caulking; the planks are fastened to the frames but not to one another, except where their ends meet at butt joints.

caulk—1) Generally, to fill any seam with a material that makes the seam watertight. 2) To fill the seams between carvel planks by pounding in rough cotton or oakum.

caulking iron—A tool shaped like a chisel with a wide, splayed, blunt edge; used with a caulking mallet to pound cotton or oakum into the seam between carvel planks.

caulking mallet—A hammer with a long, cylindrical head made of a hard wood, often mesquite; used for caulking a boat's hull.

chine—The angle where a side meets the bottom in a V- or flat-bottomed boat.

chine log—A longitudinal member running from stem to transom and notched into the chine angle of all frames, forming a watertight joint between the bottom and side planking.

clinker—A type of planking, synonymous with "clenched" and "lapstrake," in which the bottom edge of each plank overlaps the top edge of the plank below it. "Clinker" refers to fastening the overlapping planks together by driving through nails and bending the ends over, a technique called "clinking" or "clenching."

cold-molding—A process of laminating in which thin boards are glued in layers with a waterproof glue, often an epoxy, with each layer running at a right angle to the preceding layer. Usually used to fashion a boat's hull, but also used to cover decks or cabin tops. It is called cold-molding because no heat is applied to solidify the glue.

cure—With regard to the polyester resin used in fiberglassing, to completely polymerize, or harden.

double ender—A boat that tapers to a narrow backbone element (stem or very narrow transom) at both ends.

duck—A leaden weight with a bent prong extending from one end, used to hold down a spline in drawing curved lines, as in a boat design.

fair—1) Adj. referring to a curved surface or line having no visible humps, flat spots, or hollows. 2) Verb. To detect and remove such irregularities from a curved surface or line.

fairing—The process by which irregularities in a curved line or surface are detected and removed.

fairing batten—A long, flexible, straight-edged batten used for fairing.

fasten—To join together with glue and/or nails, bolts, or screws.

fiberglass—Pressed or woven cloth made of glass fibers, used with polyester resin to create a hard laminated material; colloquially refers to the entire laminate, as in "fiberglass hull." See also "FRP."

floor timber—A skeletal element fastened transversely across the lower part of a frame, just above the keel; serves to strengthen the frame and can support such elements as the engine bed.

frame—A transversely oriented member of a boat's skeleton, lying against the planking; referred to colloquially as a "rib" by laymen and some boatbuilders.

frame mold—See "mold."

FRP—Acronym for fiberglass reinforced plastic, the laminate colloquially referred to as "fiberglass."

futtock—Separate sections of a sawn frame or frame mold that are joined together with a scarf joint or gusset.

garboard—The bottom-most plank, lying closest to the keel.

guard—A longitudinal member of hard wood attached outside the upper part of a hull; protects the hull from damage in rubbing against docks or other boats.

gusset—A flat piece of wood used to join two other pieces or to strengthen the joint between them. For example, the futtocks of a sawn-frame boat are often joined together with gussets.

half-breadth plan—The part of a lines drawing that shows the hull as seen from above, including the outlines (waterlines) of horizontal sections through the hull at regular intervals above the baseline.

half-model—A scale model of one lateral half of a boat hull, carved in wood.

horn timber—As the aft-most member of the backbone, a timber supporting an overhanging stern or connecting with the base of the transom.

keel—The longest structural element of a boat's backbone, running in the midline along the bottom of the hull. Also used to refer to a fin-like external stabilizing structure attached to bottom of the keel.

kick—Colloquial term applied to catalyzed polyester resin, meaning to begin to harden or polymerize, a process that generates heat.

knee—A stout member made of wood or metal and having two arms, used to strengthen a joint, usually between backbone members; for example, small boats often have knees where the stem and transom meet the keel.

lapstrake—A type of planking in which the bottom edge of each plank (strake) overlaps the top edge of the plank below it, similarly to the way shingles overlap on a roof. See also "clinker."

lay up—To produce a fiberglass laminate by adding layer upon layer of resin-soaked fiberglass. "To lay up a hull" means to produce a hull in this manner.

lines—1) A lines drawing; 2) the functional shape of a hull.

lines drawing—The part of a designer's boat plan that gives the shape of the hull, portrayed as the outlines of sections taken at intervals through the boat in three mutually perpendicular directions; consists of the body plan (transverse sections), half-breadth plan (horizontal sections), and profile plan (vertical-longitudinal sections).

lining off—The process of marking the edges of all planks onto the frames of a carvel-planked boat prior to planking, in order to determine the widths and shapes of the planks, which taper toward the bow and stern.

lofting—Expanding the designer's lines drawing for a boat hull to full size on a whitewashed wooden floor, in order to accurately determine the shapes and bevels of the frames and other elements of the boat's skeleton, and to make full-sized patterns for those elements.

marking staff—A thin batten used in lofting to transfer heights or widths from one plan of a boat to another; marking staffs eliminate the need to make measurements with a rule and record them as numerals, both of which are time consuming and subject to error.

master bevel board—A quarter-circle with a radius of 28 and 5/8 inches, drawn on a piece of plywood or on a loft, marked with lines radiating from the center to the circumference and indicating angles from 0 to 90 degrees; used in making a bevel stick and as an aid in determining frame bevels on the loft.

mat—Abbreviated term for chopped-strand mat, a pressed cloth of fiberglass with short, irregularly oriented fibers, used alternately with roving in laminated fiberglass structures.

MEKP—Acronym for methyl ethyl ketone peroxide, the catalyst mixed with polyester resin to cause it to harden.

mold—1) In fiberglass construction, any concave, three-dimensional pattern into which fiberglass is laid to produce an object; themselves constructed of fiberglass, molds are used to produce hulls and decks, among other parts. 2) In wooden boat construction, a mold (also called a "frame mold") is a temporary transverse member having the cross-sectional shape of the hull at a particular position; a series of molds fastened at regular intervals to the backbone or to a strongback allows accurate construction of a boat's skeleton and ultimately determines the shape of the hull, after stringers, bent frames, and planking are added.

oakum—A caulking material consisting of tarred hemp or jute fibers.

offsets—Also called a table of offsets; a table of full-scale measurements taken with an architect's rule from the scaled lines in a boat's plan, indicating actual heights from a baseline and widths from a centerline. The offsets are necessary to expand the scaled lines in a boat plan to full-scale lines on the loft.

pattern—The outline shape of a piece of a boat, cut out in quarter-inch plywood and used to trace the shape onto stock lumber. A pattern in boatbuilding has the same function as a pattern in sewing.

plug—1) Noun. In fiberglass boat production, a full-sized mock-up of a boat hull or other part, used as the model from which to produce a mold, which is in turn used to fabricate multiple identical parts. 2) Noun. A short, cylindrical piece of wood with the grain running transversely rather than longitudinally, used to fill screw countersink holes. 3) Verb. To fill screw countersink holes by gluing in and later trimming plugs.

polyester resin—A liquid plastic precursor substance composed of long organic molecules not cross-linked together, used as the matrix in fiberglass laminates. The mixture sold as polyester resin for fiberglassing contains 30–50% styrene.

profile plan—The part of a lines drawing that shows the outline of a boat in side view; it includes buttock lines, which are the outlines of vertical-longitudinal sections taken at regular intervals laterally from the midline.

rabbet—A groove running the length of a backbone timber such as the stem or keel, into which will fit the ends or edges of the planking. More generally, a channel, groove, or recess cut into the edge or face of a surface.

raking transom—(See also "transom") A tilted transom. One tilted forward at the lower end is "forward raking;" one tilted forward at the upper end is "reverse raking."

real bevel—The actual bevel, or angle, that must be cut on a piece of wood to allow it to fit properly in place (see also "apparent bevel").

resin—Colloquial term for the mixture of polyester resin and styrene used in fiberglassing.

rib—Colloquial term ("the ribs of a boat") for a transversely oriented member of a boat's skeleton that lies against the outer covering of the hull; usually called a "frame" by boatbuilders.

ribband—A long, temporary batten fastened longitudinally along the frames or molds of a boat to hold these elements in place as the skeleton is set up.

roller—A hand tool something like a paint roller, but with a smaller, grooved working end made of metal; used to remove air pockets and ensure complete penetration of resin into mat and roving during fiberglassing.

roving—A heavy, coarsely woven cloth of fiberglass, used as the main structural element in laminated fiberglass structures.

Samson post—An external bitt or cleat near the bow of a boat, well attached to the skeletal structure so as to withstand heavy forces.

scantlings—The dimensions of the skeletal elements and planking in a boat. A boat with thick timbers and planking relative to its size is said to have "heavy scantlings" and one with light timbers and planking relative to its size, "light scantlings." The scantlings depend on the function for which a boat is intended; an icebreaker will need heavy scantlings and a racing boat light scantlings.

scarf—1) Verb. To join two pieces of wood together end-to-end by complementarily tapering or notching an end on each piece and gluing or otherwise fastening them together, producing a joint with no increase in cross-sectional area. 2) Noun. A joint produced by this process.

sheer—The intersection of a boat's hull and deck.

sheer clamp—A longitudinal member running inside along the tops of the frames, or notched into the outside of the frames at the sheer; in open boats, an inner rail running along the tops of the frames.

sheer line—The intersection of hull and deck along the length of a boat, seen in profile or from above, either on an actual boat or in a lines drawing.

shell construction—A mode of constructing a boat hull in which the planking is completed first to determine the shape of the hull, with frames added later inside the planking. Associated with lapstrake planking.

skeleton construction—A mode of constructing a boat hull in which the skeleton (backbone and frames) is built first, followed by planking over the frames. Usually associated with carvel planking.

slick—A very large chisel with a hefty handle, held with both hands and used for heavy-duty carving on boat timbers.

spile—To determine the shape of a plank by tacking down a long batten where the plank will go and marking on the batten the width and edge bevel of the plank at intervals; the batten is then laid down on lumber and the shape transferred so that the plank can be cut.

spline—A flexible drafting batten made of wood or plastic, about 1 meter long, with a groove along one or both edges for use with drafting ducks; used to draw smooth curves when designing a boat.

standing bevel—An angle on a piece of wood that is cut at greater than 90 degrees from a reference line.

stem—The foremost structural element of the backbone, to which the planking tapers and is attached at the bow of a boat.

stern—The rear end of a boat.

strake—1) An old term for a plank. 2) One row of the planks that comprise the covering of a hull.

stringer—A long member fastened fore-and-aft along the insides of the frames to strengthen the skeleton of a boat.

strongback—A sturdy, rigid, level frame upon which the hull of a small wooden boat is built upside down.

styrene—A clear, liquid plastic precursor substance composed of relatively small organic molecules; decreases the viscosity of polyester resin and, in the presence of a catalyst, forms cross-links with the polyester molecules, causing the mixture to become a hard plastic.

table of offsets—see "offsets."

thwart—A transversely oriented seat in a boat.

transom—The vertical or angled, flat or curved, transversely oriented structure that forms the blunt stern of a boat; in a typical outboard motorboat, for example, the transom is where the motor attaches.

under bevel—An angle on a piece of wood cut at less than 90 degrees from a reference line.

unfair—Adjective applied to a curved surface or line that has detectable humps, flat spots, or hollows.

Illustrations

Most of the photographs presented here are among the approximately 700 black-and-white images I took when I attended the Bates Boat Building program from fall 1977 to spring 1979. Around 500 of the photographs and some video clips are currently available (June 2022) for viewing and downloading via the website https://www.trumblyshow.com, although this site will not remain active indefinitely.

I have donated (2022) all my boatbuilding photographs (negatives, contact proof sheets, and a positive digital scan of each negative) to the Tacoma Public Library and transferred the copyright to that institution, which at present has a "free use" policy.

A selection of the images will continue to be available online through the ORCA (Online Records and Collections Access) web link maintained by the Library's Northwest Room (https://www.tacomalibrary.org/northwestroom/). For questions regarding access to the images or their use, please contact the Northwest Room.

Some of the following illustrations have upper and lower panels, referred to in the text with the notations **a** (above) and **b** (below), for example, **Illustrations 5a** and **9b**.

Illustration 1. Boatbuilding classroom at Bates. Lenny Viola is designing a boat, with Trumbly helping another student in the background (1978).

Illustration 2. Bates boatbuilding shop, with student Rob Cramblet (1977).

Illustration 3. Pouring a lead keel in South Tacoma (1977).

Illustration 4. Boatbuilder Ed Opheim aboard *Ragna Olsen*, Pleasant Harbor, Spruce Island, near Kodiak, Alaska (1978).

Illustration 5. Opheim dories. **Above,** dory anchored at sea, Kodiak, Alaska. **Below,** dory in a warehouse, Kodiak.

232

Illustration 6. M. Dick standing near the renovated gitney *Devil's Paw,* Spruce Cape, Kodiak (1977).

Illustration 7. Lofting floor at Knapp Boatbuilding / CLK Yacht Crafters, with a completed half-frame pattern for Aquila (1978).

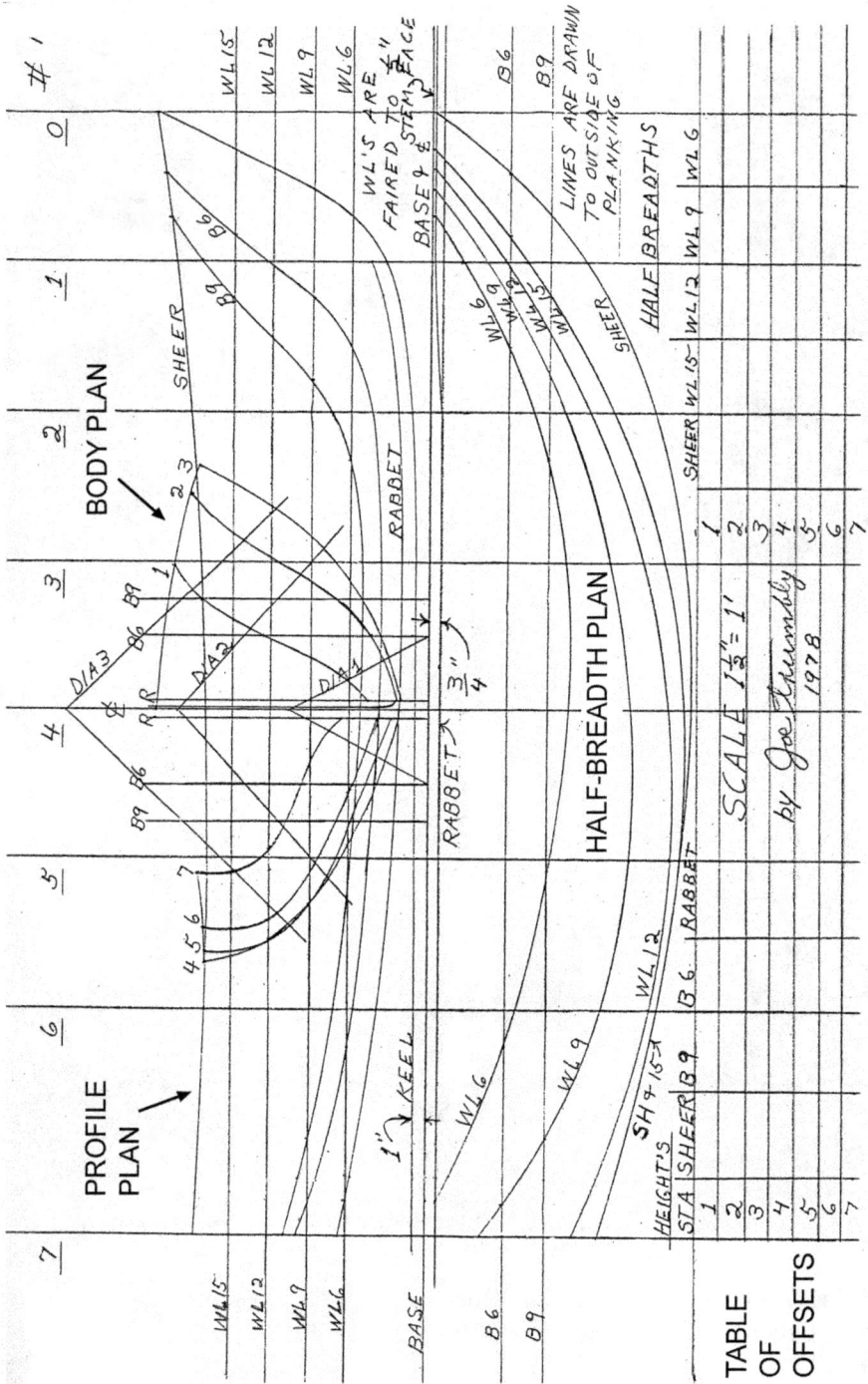

Illustration 8. Lines for Trumbly's lofting exercise (1978).

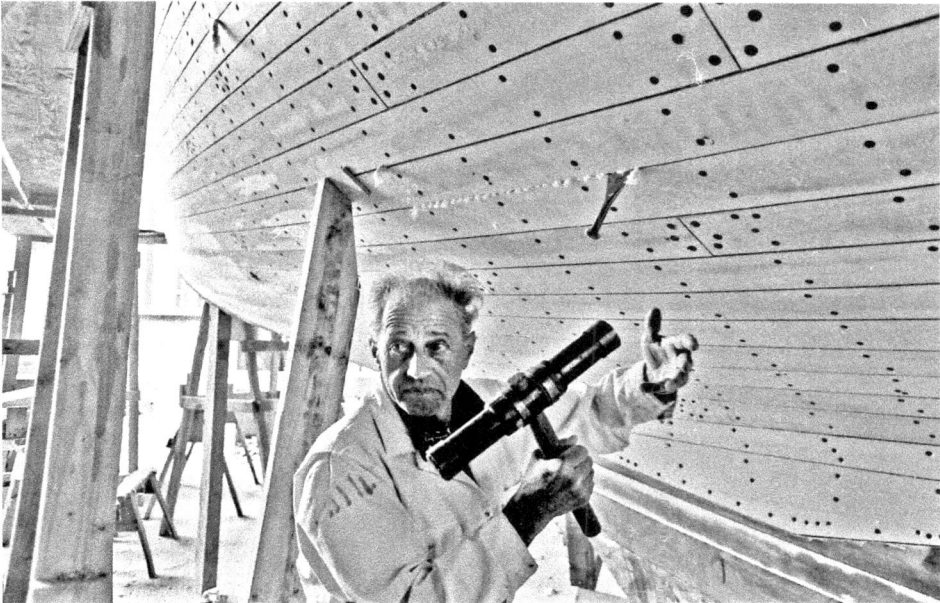

Illustration 9. Skeleton construction. **Above,** Brian Saucier's 19-foot boat, with stringers in place and many of the bent frames installed. The temporary frame molds can be seen inside the boat's skeleton (Bates shop on the South Campus, 2003). **Below,** carvel planking, associated with skeleton construction; Trumbly is wielding a caulking mallet, demonstrating caulking on the T-38 (1978).

Illustration 10. Loft in the Bates boatbuilding shop, with Dean Goodrich in the foreground working on Trumbly's lofting exercise and another student far in the background working on the main lofting floor (1977).

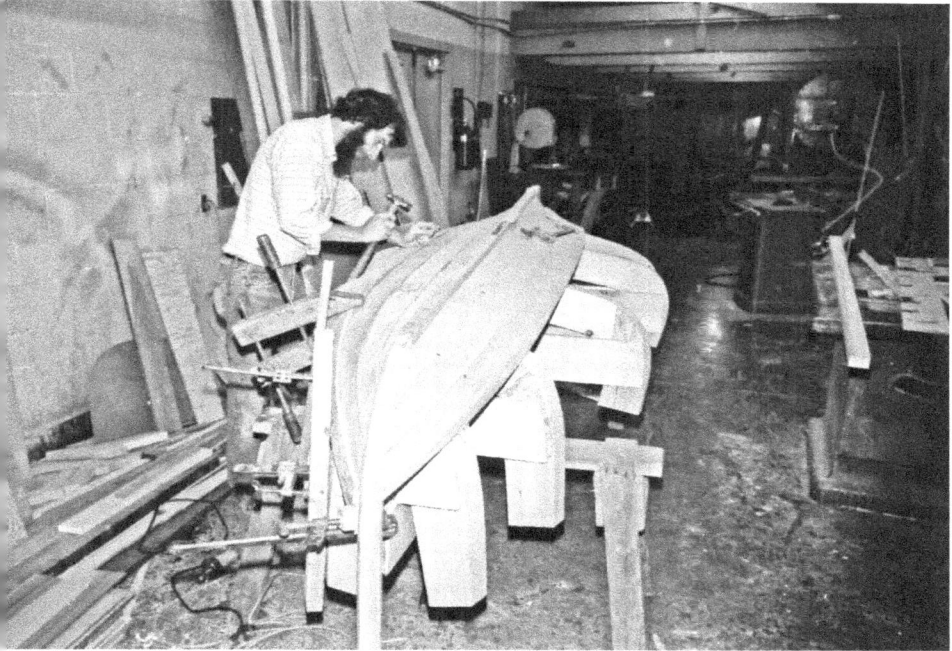

Illustration 11. Student building a lapstrake boat at Bates. Although lapstrake boats were traditionally built by using shell contruction, the mode being used here is skeleton construction (1977).

Illustration 12. Above, Patrick Chapman on the bow of the T-38, preparing to install Samson posts (1977). **Below,** cabin top under construction (1978).

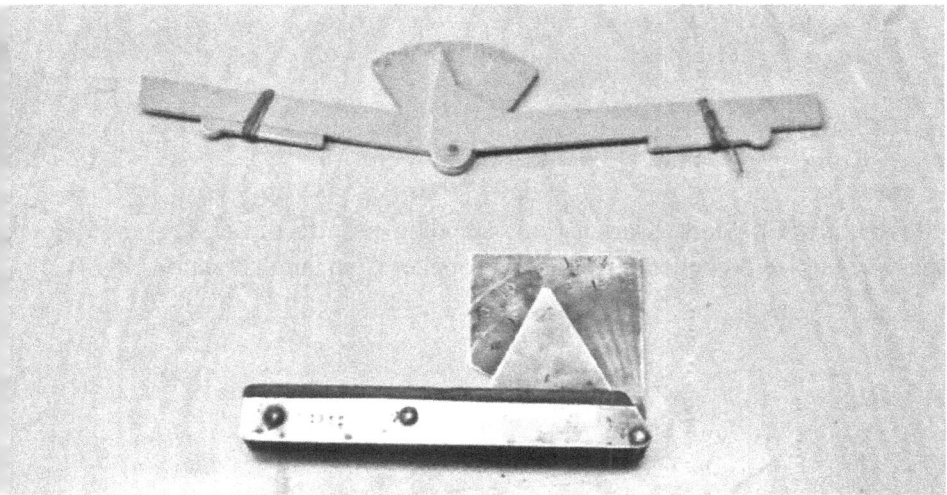

Illustration 13. Above, Toolroom duty at Bates (1977). **Below,** Trumbly-designed bevel gauges typically made by students at Bates using Trumbly's patterns, a self-reading gauge for plank bevels (upper) and a self-reading gauge for any bevel (lower).

Illustration 14. Mark Klarich (left) standing next to the casting mold for a massive bronze propeller at Coolidge Propeller Company in Seattle (1978).

Illustration 15. Trumbly's propeller-design display, showing stages from design on paper through half-blade patterns (left and lower left, dark pieces), whole-blade sections, unfinished laminated wooden models, finished wooden models for casting, and finished cast bronze propellers (right).

242

Illustration 16. Trumbly on a sawhorse, checking the fairness of a batten while lining off a sailboat at Evergreen College (1977).

Illustration 17. Planking the T-38. **Above,** newly installed garboard plank, with a spiling batten for the next plank up; note the plywood pattern in place for the forward end of the plank being spiled. **Below,** variable plank shapes above the garboard (1978).

Illustration 18. Tilting-arbor plank cutter for cutting a variable bevel along the length of a plank, said to have been used at Tacoma Boat in the days of production wooden boatbuilding.

Illustration 19. Trumbly's 40-foot ketch *Osage*. **Above,** under sail; Trumbly steering, with Patrick Chapman standing behind him. **Below,** tied up in Gig Harbor (1977).

Illustration 20. George Chambers setting up to drill the rudder shaft hole on the T-38 (1978).

Illustration 21. Don Knapp standing near the stern of *Macs' Effort* during propeller installation (1978).

248

Illustration 22. *Macs' Effort* in the Knapp Boatbuilding shop. **Above,** view of the deck, with a worker adding finishing touches. **Below,** just prior to launching, with (L to R) Don Knapp and George and Bob McPherson (1978).

Illustration 23. Trumbly-designed fiberglass sailboat mold and hull at the Seawind Company in Puyallup (1978).

Illustration 24. Bates boatbuilding class setting up the backbone for Trumbly's 51-foot sailboat *Windance* on Raft Island, with Trumbly hamming it up in the foreground (1978).

Illustration 25. Field trips. **Above,** Bates students examining the ULDB (ultra-light-displacement boat) *Merlin* undergoing repairs at Eddon Boat in Gig Harbor (1978). **Below,** the Ron Jones hydroplane shop in Seattle (1978).

252

Illustration 26. M. Dick's 14-foot sailboat under construction at home (1978).

Illustration 27. Aquila. **Above,** Griz setting up frame molds on a strongback at CLK Yacht Crafters. **Below,** all frame molds set up, with Chuck Krapp (left) and Earl Edwards (1978).

Illustration 28. Aquila plug at CLK Yacht Crafters. **Above,** sanding (1978). **Below,** being extracted from the shop with a crane after removal from the mold; the plug will be discarded (1979). Bottom photograph courtesy of Chuck Knapp.

Illustration 29. Applying the balsa core to the mold for Aquila at CLK Yacht Crafters (1978).

256

Illustration 30. Seattle Boat Show in the King Dome (1979).

Illustration 31. Perk Haynes at Hylebos Boat Haven, standing near a dory built by a former Bates student (1978).

Illustration 32. Dory under construction in Perk Haynes's garage (1979).

Illustration 33. Doug White (left) and Steve Thrasher, with the restored *Adelphi* in the background (1979).

Illustration 34. M. Dick standing on the deck of *Adelphi*, with Patrick Chapman in the background working on bulwark installation (1979).

Illustration 35. Dan Hubley plugging countersink holes on the T-38 (1978).

Illustration 36. Shaft-boring jig being used to bore the propeller shaft hole for *Windance*. View from inside the keel (1979).

Illustration 37. Shaft-boring jig being used to bore the propeller shaft hole for *Windance*. Two views from outside the keel. At the bottom of the top panel is Mark Klarich with a dog (1979).

Illustration 38. *Adelphi*'s keyhole rudder (1979).

Illustration 39. Mark Klarich standing next to the restored *Adelphi* on launching day (1979).

Illustration 40. Patrick Chapman with his nearly finished dory shortly before he left Bates, standing with Joe Trumbly (1979).

Illustration 41. Etta Trumbly and the A-frame home she and Joe built on Raft Island (1978). **Above,** Etta watching the boatbuilding class setting up *Windance*'s backbone. **Below,** Etta standing on the deck.

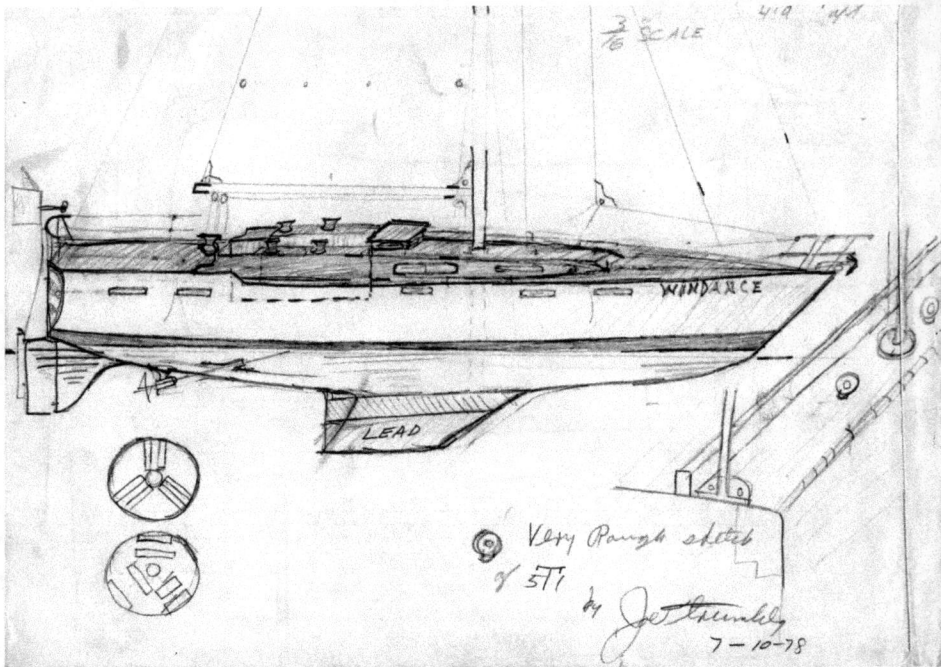

Illustration 42. Drawing of *Windance* by Joe Trumbly (1978).

Illustration 43. Joe Trumbly sitting inside the skeleton of *Windance,* under construction on Raft Island (1979).

Illustration 44. M. Dick's 19-foot pram-bow skiff *New Valaam*, constructed and launched in Ouzinkie, Alaska (1980).

Illustration 45. Boat Building shop in the new facility on the South Campus of Bates Technical College, formerly LH Bates Vocational Technical Institute. For scale, a full-sized replica of the famous 24-foot sailboat *Seraffyn* (with a white hull) can be seen near the back left corner (2003).

Illustration 46. Seven and one-half foot boat under construction from the lines for Trumbly's lofting exercise shown in Illustration 8, undertaken by students in the last years of Bates Boat Building (2003).